ID0597674

LAN to WAN interconnection

LAN to WAN interconnection

John Enck
Mel Beckman

McGraw-Hill, Inc.

New York San Francisco Washington, D.C. Auckland Bogotá
Caracas Lisbon London Madrid Mexico City Milan
Montreal New Delhi San Juan Singapore
Sydney Tokyo Toronto

©1995 by **McGraw-Hill, Inc.**

Printed in the United States of America. All rights reserved. The publisher takes no responsibility for the use of any materials or methods described in this book, nor for the products thereof.

hc 1 2 3 4 5 6 7 8 9 DOC/DOC 9 9 8 7 6 5

Product or brand names used in this book may be trade names or trademarks. Where we believe that there may be proprietary claims to such trade names or trademarks, the name has been used with an initial capital or it has been capitalized in the style used by the name claimant. Regardless of the capitalization used, all such names have been used in an editorial manner without any intent to convey endorsement of or other affiliation with the name claimant. Neither the author nor the publisher intends to express any judgment as to the validity or legal status of any such proprietary claims.

Library of Congress Cataloging-in-Publication Data
Enck, John, 1956-
    LAN to WAN interconnection / by John Enck, Mel Beckman.
      p. cm.
    Includes index.
    ISBN 0-07-019614-1
    1. Local area networks (Computer networks)   2. Wide area networks
(Computer networks)   I. Beckman, Mel, 1955- . II. Title.
TK5105.7.E53 1995
004.6'8—dc20   95-10086
                                                                      CIP

Acquisitions editor: Marjorie Spencer
Editorial team: Marc Damashek, Book Editor
               David M. McCandless, Managing Editor
               Joanne Slike, Executive Editor
Production team: Katherine G. Brown, Director
               Rhonda E. Baker, Coding
               Brenda M. Plasterer, Coding
               Wanda S. Ditch, Desktop Operator
               Nancy K. Mickley, Proofreading
               Brenda S. Wilhide, Computer Artist
               Joann Woy, Indexer
Design team: Jaclyn J. Boone, Designer   0196141
               Katherine Stefanski, Associate Designer   EL3

In memory of my father, Herbert,
my mother, Birdine,
and Taz, my devil dog.
John Enck

To my wife, Patricia
Mel Beckman

Acknowledgments

The authors would like to thank Duke Communications for their encouragement and support over the years; Norm Bartek for his careful and skillful review of our first draft; Marjorie Spencer, for her patience and guidance during the development of this book; and McGraw-Hill for the chance and reality of bringing this book to market.

Contents

Part III: Interconnection tools

Part IV: High-speed interconnections

Preface

One of the most important issues we discussed before we began writing this book was how technical the material should get. For example, did we want to discuss the properties of electrons moving through copper media? Did we want to explain the principles that allow light waves to travel through bends in fiber optic cable? Did we want to include lengthy dissertations on the differences between Manchester encoding and differential Manchester encoding?

After some brief discussions along these lines, we finally agreed that although these principles are certainly important, they aren't particularly relevant when it comes to the day-to-day reality of interconnecting networks. Having made that decision, we then agreed to avoid the use of academic terminology in favor of common terms and terminology. Our goal was simple: we wanted to create a no-nonsense book that presents the information you need to know to plan, implement, and manage Local Area Network (LAN) connections, Wide Area Network (WAN) connections, and, of course, LAN-to-WAN interconnections.

How successful were we in reaching our goal? That decision is yours to make, and we encourage you to do so.

John Enck
Mel Beckman

Introduction

This book is divided into five parts. Each part begins with an introduction that provides an overview of the technology discussed in the related chapters. The chapters following that introduction then provide detailed information on technology. The specific organization of the parts and chapters is as follows:

Part I: General overview. A brief discussion of the basic LAN/WAN problem.

Chapter 1: When LANs meet WANs. The purpose and application of Local Area Network (LAN), Metropolitan Area Network (MAN), and Wide Area Network (WAN) technology.

Part II: LAN basics. Information critical to the planning, installation, and expansion of Ethernet/IEEE 802.3 and Token Ring/IEEE 802.5 LANs.

Chapter 2: Ethernet/802.3. Detailed information on Ethernet and IEEE 802.3 LANs, including discussions on topology, construction, and the common Carrier Sense, Multiple Access with Collision Detection (CSMA/CD) access discipline. Also discusses high-speed Ethernet technology.

Chapter 3: Token Ring/802.5. Detailed information on IBM Token Ring and IEEE 802.5 LANs, including discussions on topology, construction, and the common token-passing access discipline. Also discusses the pros and cons of source routing.

Part III: Interconnection tools. Examination of the fundamental tools that interconnect multiple LANs in a local or wide area environment.

Chapter 4: Bridges & routers. An examination of bridging and routing technology and applications. Also includes definitions of the major bridging and routing protocols and techniques, e.g., the Routing Information Protocol (RIP), the Open Shortest Path First (OSPF) protocol, and Data Link Switching (DLSw).

Chapter 5: Gateways & hubs. An examination of gateway and hub functions and applications. Also includes discussions on protocol encapsulation, how gateways operate in Transmission Control Protocol/Internet Protocol (TCP/IP) networks, and "smart" versus "dumb" hubs.

Part IV: Specialized high-speed interconnections. Explorations of high-speed interconnections that can be used to implement LANs, MANs, or interconnect LANs/MANs to create a WAN.

Chapter 6: Fiber links & FDDI. An exploration of the networking applications for fiber optic links. Includes discussion of how fiber optic links operate, and of the Fiber Distributed Data Interface (FDDI).

Chapter 7: Frame & cell relay. An exploration of frame relay technology, cell relay technology, and the Asynchronous Transfer Mode (ATM) architecture.

Part V: Traditional WAN interconnections. Information on conventional wide area connections commonly used for low-speed network connections (less than 100 Mbps).

Chapter 8: Point-to-point links. A look at switched and leased analog phone links, digital phone links, and T1, fractional T1, and T3 services.

Chapter 9: Multi-point links. A look at X.25 packet switching networks, Integrated Services Digital Network (ISDN), and Switched Multimegabit Digital Services (SMDS).

General overview

IF YOU THINK ABOUT IT, NETWORKING TECHNOLOGY HAS become so intertwined with computer technology that it is often difficult to separate the two. For example, users stationed at terminals depend on the link between their terminals and the main computer just as much as they depend on the computer itself. Similarly, PC users who access files and printers on a common server rely on the network that interconnects them with the server. And in many large organizations, terminals, host computers, PCs, and PC servers rely on network connections to perform day-to-day business transactions—even though these devices may be scattered across the face of the globe.

Like computers, network technology comes in all shapes and sizes. Simple network connections accommodate the physical interface between a PC and a modem, or a PC and a printer. More intricate connections handle the attachment of multiple terminals to a common host, or facilitate the interconnection of PCs and PC servers. As you progress farther up the scale of network technology, you run into complex connections that function as the building blocks for large (possibly world-wide) networks.

All of these types of connections—from the lowly PC serial cable up to world-spanning fiber links—are important components in the grand scheme of networking. Unfortunately, they cover so much technical ground that it is difficult to get your arms around networking as a whole. Consider, for instance, how vastly different the electrical current technology used to carry information in most Local Area Networks (LANs) is from the light-wave technology used to carry information over fiber optic links.

Even the application of networking technology varies from one situation to another. The networking solution that works well for interconnecting PCs in a single office building may fall flat on its face when the network expands into multiple buildings located miles apart. The wide area strategy linking remote terminals back to a central computer may crumble when the terminals are replaced by PC LANs. Any network solution appropriate for a specific computing environment is subject to change if that environment changes.

As complicated as networking is, it is not rocket science. You do not need a degree in electrical engineering or quantum physics to understand and appreciate its subtleties. If you have a good working understanding of one aspect of networking, you'll find it relatively easy to learn other aspects. For example, if you've been working with LANs for several years, you'll find that Wide Area Network (WAN) technology has many familiar elements. Or if you've been working with wide area links for years, you'll find similarities in how local area links operate.

And what if you haven't been working with either LAN or WAN technology? Don't worry—we make no assumptions about your skill level. In fact, we cover a broad spectrum of networking technology that includes LANs, WANs, and the overlap between the two. We hope that this approach will save wear and tear on two of your most highly treasured networking tools: your time and patience.

When LANs meet WANs

NETWORKING HAS BEEN A KEY ELEMENT IN DATA PROCESS-ing since it moved from a batch environment to an interactive environment in the early 1960s. With that transition, the computer input/output environment changed from card readers, paper tape readers, and line printers to remote terminals and dot matrix printers. Programmers, operators, and eventually users learned to interact directly with the computer—punched cards and paper tape strips were no longer required.

The creation of an interactive user environment was quickly followed by the introduction of computer-to-computer links to allow computers to exchange information with one another. At first these links required the computers to be physically close to one other, but over time the technology matured to the point where computers could communicate with each other regardless of distance. Once the interconnection technology was established, the focus turned to speed—how to move information over these computer-to-computer links faster and more efficiently.

Taken together, these two developments—interactive users and computer-to-computer links—formed the foundation for today's towering empire of Local Area Network (LAN) and Wide Area Network (WAN) technology.

Despite this common foundation, LAN and WAN technology evolved separately. WAN technology was developed to address the requirements of a centralized computer network, while LAN technology was advanced to handle the issues associated with a distributed computer network. This distinction can best be seen by looking at the network approaches used by two major computer companies—IBM's centralized network architecture and Digital Equipment's distributed network architecture.

The need for wide area networking

Under IBM's vision for computing in the 1960s and 1970s, a large central computer (i.e., a mainframe) provided processing and mass storage services to handle both interactive and batch applications. This central computer was designed to provide application services to hundreds (and eventually thousands) of remote terminals and printers. Given this scope of operation, it became clear to IBM that a single central computer could not efficiently handle both the application load and the communications load.

IBM's solution to keeping communications overhead off the central computer was to develop a hierarchical network structure that distributed the communications functions over a number of devices, as shown in Fig. 1-1.

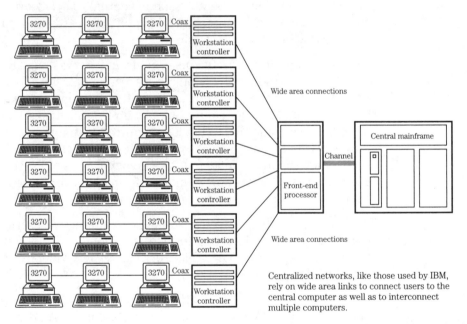

Centralized networks, like those used by IBM, rely on wide area links to connect users to the central computer as well as to interconnect multiple computers.

Shown here is a typical IBM terminal network.

■ **1-1** *Centralized network implementation*

Under this design:

☐ User terminals and printers attach to workstation controllers using coaxial connections. A single workstation controller handles a number of terminals and printers.

☐ Each workstation controller is responsible for delivering messages sent from the central computer to its attached terminals and printers, and for collecting information entered into terminals and transmitting it to the central computer. Workstation controllers can interface with either the central computer, or more likely, with a communications device called a "front-end processor" (FEP).

☐ The purpose of a FEP is to further isolate the central computer from the communications network. The FEP manages the traffic to and from workstation controllers and then funnels it into the central computer over a specialized high speed connection called a "channel" attachment. This is the same kind of attachment used to connect disk drives and tape drives to the central computer.

As you can see, IBM's hierarchical design frees the central computer from handling the communications environment, because communications functions are distributed across the FEP and workstation controllers. IBM continued to refine its network design, adding capabilities like FEP-to-FEP communications so multiple central computers could communicate with one another, and could share terminal and printer resources as well. In 1974, IBM turned their network design into a formal architecture, called the Systems Network Architecture (SNA).

IBM's centralized network design depends on wide area links to work effectively. For example, if a group of users located in Boston needs to communicate with a central computer located in New York, a wide area link is required to connect a workstation controller in Boston with the FEP in New York. Or if a central computer in Los Angeles needs to exchange information with a sister company's central computer located in Dallas, then a wide area link needs to be put in place between the two FEPs.

In short, when you introduce significant geographical distances into a centralized network, that network's dependency on wide area technology increases.

Now IBM was not the only company with a network design dependent on wide area technology—many other major computer manufacturers followed IBM's lead and introduced centralized network designs. By the time the 1980s rolled around, the majority of the computer industry was clamoring for wide area technology to improve the speed and efficiency of their centralized network designs.

Wide area network evolution

The first major milestone for wide area technology was the creation of a modulation–demodulation device, better known as a "modem." A modem translates digital information sent from a computer into an analog format suitable for transmission over a standard telephone line. A second modem on the other end of the line then translates the analog information back into digital information for the other computer. Because telephone lines are available over a wide geographical area, they were (and still are) perceived as an ideal means of transmitting terminal, printer, and other computer information over long distances.

In the beginning, modems supported very modest transmission speeds—for example, 300 bits per second (bps). Over time, modem technology improved and transmission speeds gradually increased. Thus the transmission rate of 300 bps was superseded by 1200 bps, then by 2400 bps, 4800 bps, 9800 bps, and so on, until today's current speed of 28.8 kbps (28,800 bps).

The way that modems interacted with the phone lines also changed over time. In the beginning, modems could work over standard voice phone lines or dedicated leased lines. When standard voice lines were originally used, someone (some person) had to manually dial up the other modem to establish the connection. This manual intervention was viewed as less than ideal, so the Automatic Calling Unit (ACU) was developed to handle the dialing automatically. The ACU technology was then surpassed by the Hayes modem command set, which remains the de facto standard for modems today.

Leased phone lines avoid the issue of dialing because each end of the line is permanently assigned to a modem. Leased lines, however, brought their own set of technology issues. For example, leased lines can be configured for two wires or four wires, and can even be configured so that a single line has multiple connections, allowing a pool of modems to share the line. The dedicated nature of leased lines is very attractive for centralized network implementations—wide area links can be available 24 hours a day, 7 days a week, with no chance of accidental hang-ups.

Unfortunately, the analog technology used in voice telephone lines has its limitations. The most important limitation is that because the computer information is sent in an analog format, any "noise" or disturbances on the line have a significant negative impact on the transmissions. So the long distance phone carriers created and offered another option for computer networking—digital links.

Unlike analog phone links, digital links do not need to have the computer's digital information translated into an analog format—instead, the digital information is sent across the wide area link in its native form. This approach eliminates the need for a modem; however, a physical device is still needed to connect the computer to the digital line. This device is called a Data Service Unit (DSU).

Digital links are available in a variety of packages. For example, you can buy a simple Digital Data Service (DDS) link to go between two points at transmission speeds of 56 kbps. Alternatively, you can buy "T1" or "T3" links that contain multiple 64 kbps links offering a total aggregate speed of 1.544 Mbps (1,544,000 bps) and 45 Mbps (45,000,000 bps), respectively. Other packages are also available.

The biggest advantage of digital links is the transmission speed they offer. For example, a "low speed" digital link offers a speed of 56 kbps, a rate much higher than the best possible speed of 28.8 kbps over analog phone lines. The biggest disadvantage of digital links is that, like leased analog lines, they require a dedicated connection. You cannot dial into a digital circuit from just anywhere—you need a fixed point of entry.

Limited access to digital links is one of the issues that drove phone carriers to the development of the Integrated Services Digital Network (ISDN). ISDN offers the speed advantages of digital links while still offering the ability to "dial" other systems in the network. This dial capability is conceptually similar to the way that a computer can dial other computers over standard voice lines.

ISDN is based on digital links similar to the ones used in T1 connections, but ISDN is different from other digital services in that it is intended as a solution for both voice and computer information. Under the grand vision for ISDN, the existing analog phone lines will slowly be phased out of existence in favor of digital ISDN links.

Unfortunately, ISDN has suffered some major setbacks. For one, the deployment of ISDN links has been slow—ISDN services are still not available in many areas of the United States today, and international usage is far short of US usage. This shortage has led to other short-term technology solutions such as Switched Multimegabit Data Service (SMDS), a service offered by many local phone companies.

Another perceived shortcoming of ISDN is its speed. Although ISDN can offer up to 1,544,000 bps of transmission speed per link, this speed is no longer viewed as optimal for today's networks. Instead, high speed fiber optic networks are viewed as a better solution. This technology will be discussed in "The LAN/WAN Conflict" section of this chapter.

Thus far, this discussion on wide area technology has focused on the solutions provided by the phone carriers. Although the phone carriers own most of the physical phone wire that interconnects our homes, offices, and public facilities, they are not the only game in town when it comes to wide area networking. There are both alternative technologies and alternative carrier offerings for implementing wide area networks. The alternative technologies include the following:

☐ Satellite-based transmission. Using this technology, data transmission is bounced off an orbiting satellite to reach remote locations. As you can imagine, this is not a cheap solution.

☐ Wireless transmission. Cellular and radio wave transmissions can also be used to send or relay data transmissions over wide areas. The use of wireless transmission as a wide area solution has risen dramatically in the 1990s.

In terms of alternative carriers, these companies tend to combine various wide area technologies and offer them to their customers as a single seamless solution. For example, IBM, Digital, and Hewlett-Packard all offer "value-added" networks that include traditional phone carrier links, satellite links, and wireless links. One of the biggest advantages of using an alternative carrier is that they often offer a solution that allows you to access a number of systems from a single point of entry, without relying on ISDN links. This is often accomplished through the use of "packet switching" technology.

Packet switching networks have been around for a long time. With the exception of the CompuServe network, packet switching networks never achieved mainstream popularity in the United States. Everywhere else in the world, however, packet switching technology is an integral part of data networks.

The premise of packet switching technology is simple. Instead of creating point-to-point links (as in the case of traditional leased phone lines), packet switching networks provide a mesh of connections. In a packet switching network, any connection into the network can potentially allow you to connect to any service in the network (providing you are authorized to do so). This is dramatically different from traditional US phone carrier links (except for ISDN links) that require you to plan each potential connection in advance.

Another significant difference between packet switching networks and point-to-point or multi-point networks is the way that information is transmitted. In point-to-point and multi-point networks, data is transmitted based on the needs of the computer connection. If the computer needs to receive every character a user types on his or her terminal, then each character is sent over the link. Alternatively, if the computer wants to receive information in blocks, then the terminal (or workstation controller) sends the information as groups of characters.

In packet switching networks, however, data is organized into blocks of information called "packets." The maximum size of a packet (often 128 characters) is a function of the network itself, not the computers. Messages larger than the packet size are broken down into a series of packets. These packets are then transmitted through the network using dynamic routing that determines the best path for any given message at a given time. Thus, if a terminal transmits two messages to a host computer over a packet switching network, it is possible that the two messages will take different paths to reach the destination.

Virtually any type of link can be used inside a packet switching network—leased analog lines, digital lines, satellite links, and so forth. Similarly, the external attachment between a packet switching network and a customer can also be virtually any type of connection. In most cases, however, the external connection is a leased or dial-in connection over a standard analog phone line.

Unfortunately, using an analog phone connection has the side effect of limiting the throughput of a connection, because you cannot send or receive data any faster than the slowest link within the connection. Furthermore, many packet switching networks use internal analog connections, external analog connections, or both, so packet switching networks are often perceived as a "low speed" network solution. As the availability of high speed switched networks such as ISDN increases, the popularity and use of packet switching networks decreases.

Popularity notwithstanding, packet switching technology is a cornerstone for many of today's high speed network technologies. For example, ISDN's capability of handing multiple connections is patterned after packet switching networks. The cell-based structure of the Asynchronous Transfer Mode (ATM—further discussed in "The LAN/WAN Conflict" section of this chapter) is derived from the packet-based structure of packet switching networks. So even as packet switching networks are slowly fading from the modern networking scene, the underlying concepts of packet switching live on in today's networks.

The need for local area networking

When Digital Equipment Corporation designed its line of computer products, it wanted to offer processing and storage capabilities similar to the IBM mainframe model, but Digital did not want to copy the monolithic, centralized design of the mainframe and its attached network. Instead, Digital reasoned that if multiple computers could somehow be interconnected, the combined processing and storage capacity of those computers could effectively compete with the mainframe.

To accomplish this task, Digital designed a distributed network which allowed computers to share files, printers, and terminals with one another. Digital named its design the Digital Network Architecture (DNA), but it is better known today as "DECnet." Digital's design, shown in Fig. 1-2, differs from the centralized IBM vision in several key areas:

☐ Rising needs for processing and storage resources are met by deploying additional computers under the Digital design. Under the IBM design these same needs would be addressed by upgrading the performance or capacity of the central computer.

☐ Terminals are not attached to a specific host in the Digital network—instead, terminals are attached to "terminal servers" that allow each terminal user to attach to any host in the network. In an IBM network, terminals are attached to a central computer in a controlled and hierarchical fashion.

☐ Because the computer is dependent on the network in the Digital design, support for network routines is integrated into the computer operating system. In the IBM design, networking is kept separate and distant from the central computer's operating system.

Digital released these distributed capabilities slowly, beginning with basic file transfer capabilities, then adding remote file access and network management tools. In these initial implementations, the physical network was composed of serial links which offered limited performance. Digital quickly realized that in order for its vision of distributed computing to succeed, it required high speed interconnections.

To achieve the speed it needed, Digital had to trade off distance. Instead of having Digital computers scattered around the world and interconnected via wide area links (as in the case of the IBM network solution), Digital turned to the emerging Ethernet Local

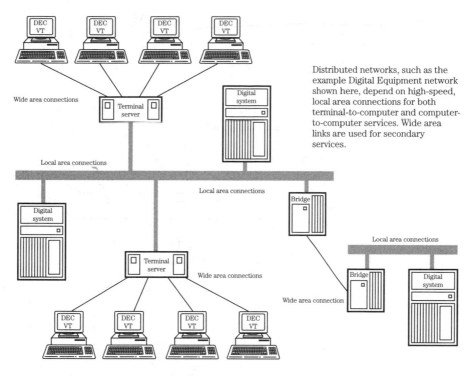

Distributed networks, such as the example Digital Equipment network shown here, depend on high-speed, local area connections for both terminal-to-computer and computer-to-computer services. Wide area links are used for secondary services.

■ **1-2** *Distributed network implementation*

Area Network (LAN) technology. Ethernet allowed Digital to interconnect its computers and terminal servers at the transmission speed of 10 Mbps. At that speed Digital's distributed services performed quite well.

Even though Digital's Ethernet-based networks were highly successful, Digital still had to address wide area connectivity under certain circumstances. Users located in remote areas might need to access the Ethernet-based Digital computers, or perhaps multiple Digital networks might need to be interconnected. To its credit, Digital addressed these needs for wide area connectivity without violating their distributed design:

☐ Because the interface between the user terminals and LAN-attached terminal servers uses simple serial connections, remote users can access a terminal server over analog phone lines (dialed or leased). Once connected to the terminal server, a terminal can potentially access any computer in the LAN.

☐ Instead of implementing computer-to-computer wide area links (as in the case of the IBM design), Digital implemented network-to-network links using bridges and routers. These devices allow information to move between high speed LANs over low speed links without adversely effecting LAN performance. This technology is also an important tool in resolving the LAN/WAN conflict, as discussed later in this chapter.

Digital's decision to use a LAN as the primary network transport, augmented by secondary wide area links as needed, proved to be both successful and popular. In all fairness, however, Digital was not the only organization investing in LAN technology. Unix vendors, such as Sun Microsystems, HP/Apollo, and others found that the Transmission Control Protocol/Internet Protocol (TCP/IP) ran very effectively in a Local Area Network augmented by wide area links. Similarly, when Xerox, Microsoft, Novell, and other companies began introducing PC-based networking products, they also turned to LAN technology to achieve the performance they required.

Local area network evolution

The early development of LAN technology dates back to the late 1960s and early 1970s; however, this technology was not actually applied to computer networks until the late 1970s. For example, the design for token-ring networking dates back to 1969, but the design was applied to phone circuits, not networks. Similarly, Ethernet technology can be traced backed to 1972, but it was used by Xerox Corporation as an internal bus for copiers.

In fact, the first major commercial computer LAN was not Ethernet or Token Ring—it was the Attached Resource Computer Network, better known as "ARCnet." Released in 1977 by Datapoint Corporation, ARCnet provided a means of interconnecting Datapoint computers so they could share files, printers, and other resources. When the PC explosion occurred in the early 1980s, ARCnet crossed over into the PC world to serve as the LAN of choice for the newly emerging file and print server market.

ARCnet uses a token-passing access discipline operating over a bus or a star topology. The transmission rate over an ARCnet network is 2,500,000 bps (2.5 Mbps), which at the time was quite respectable. Furthermore, ARCnet products were easy to develop and inexpensive to produce, market, and install, so ARCnet quickly gained the lion's share of the PC LAN market. Unfortunately for ARCnet lovers, it was a short-lived position, because while ARCnet was being deployed in one PC LAN after another, Ethernet and token-ring technology were undergoing heavy construction.

12

As noted, Ethernet was created by Xerox Corporation in 1972, but the first commercial implementation of Ethernet was developed by the joint efforts of Xerox, Digital Equipment Corporation, and Intel Corporation. The resulting product, now referred to as Ethernet Version 1, was introduced in 1980.

Although Ethernet Version 1 was successful, it was not wildly successful. One of the problems with Ethernet Version 1 was that, because it was a commercial product, there were no impartial organizations to interpret the specifications in the areas where they were less than crystal clear. Therefore each commercial Ethernet Version 1 engineer came to his or her own conclusions and implemented them. As a result, many commercial Ethernet Version 1 products turned out to be incompatible with one another.

In an effort to resolve these incompatibilities, the Institute of Electronics and Electrical Engineers (IEEE) decided to step in and define LAN standards of its own. Instead of creating its own standards from scratch, the IEEE used the technology currently available on the market as a starting place. Thus the IEEE looked at ARCnet, looked at Ethernet Version 1, and listened to IBM when it presented a series of papers on token-ring technology.

The results of the IEEE effort are the 802 series of LAN specifications. The IEEE first developed an overall access method—applicable to all types of LANs—and then developed specifications for each type of LAN. The resulting series of specifications were released in the early 1980s and included

☐ the IEEE 802.2 specification for a global LAN access methodology. This specification applies to each of the specific LAN implementations (802.3 through 802.5).
☐ the IEEE 802.3 specification for a Carrier Sense, Multiple Access with Collision Detection (CSMA/CD) LAN. This is the IEEE version of Ethernet.
☐ the IEEE 802.4 specification for a token-passing bus LAN. This is the IEEE version of ARCnet.
☐ the IEEE 802.5 specification for a token-passing ring LAN. This is the IEEE version of IBM's Token Ring LAN.

Digital and IBM both continued to offer their own commercial LAN implementations independent of the IEEE specifications. Digital came out with Ethernet Version 2, which provided coexistence (but not compatibility) with the IEEE 802.3 specification. Similarly, IBM continued to offer its own implementation of token-ring technology, although the IBM specifications are extremely close to the IEEE 802.5 specifications.

IBM's commitment to Token Ring LANs and Digital's commitment to Ethernet was the beginning of the end of ARCnet. Although it is difficult to point to a single event that pushed ARCnet off the ledge of a building, the three significant developments that helped ease ARCnet into the hereafter were the following:

☐ Ethernet and Token Ring offered higher transmission speeds than ARCnet (Ethernet clocked in at 10 Mbps, while Token Ring started at 4 Mbps and expanded to 16 Mbps).

☐ No major computer corporation embraced ARCnet as a LAN standard—ARCnet remained a "fringe" PC LAN solution. All of the major computer corporations endorsed Ethernet/802.3, token-ring/802.5, or both.

☐ For years ARCnet offered a significant price/performance advantage over Token Ring and Ethernet. Once Ethernet became a popular LAN standard in the larger corporate markets, the cost of Ethernet equipment fell to the point where Ethernet price/performance was superior to ARCnet price performance.

Although ARCnet networks remain in operation, Ethernet/802.3 and token-ring/802.5 networks far outnumber them. Today the battle is between Ethernet/802.3 and token-ring/802.5 technologies.

At the time of their introduction, Ethernet/802.3 and token-ring/ 802.5 networks clearly offered better performance than traditional wide area links. After all, even a "slow" token-ring network operating at 4 Mbps offers over 70 times the performance of a 56 kbps digital link. This high speed advantage attracted new applications, new systems, and new protocols to the LAN environment.

Before long, the performance of many LANs began to suffer due to a dramatic rise in network traffic. Higher transmission speeds were needed to support ever-increasing loads. More connections were needed to meet increasing demands for access. In short, LAN technology had to undergo further evolution.

The quest for higher performance LANs took many directions. Fiber optic backbones were deployed. Switching hubs were introduced. New signaling techniques were developed to achieve high transmission speeds over Unshielded Twisted Pair (UTP) cable. Of all the techniques, the two most significant LAN technology developments thus far have been

- [ ] the introduction of the Fiber Distributed Data Interface (FDDI) network. FDDI is an all-fiber network that operates at speeds of 100 Mbps. In addition to offering high speed transmission, a FDDI network can operate over distances of many kilometers, and is therefore often referred to as a Metropolitan Area Network (MAN) instead of LAN.
- [ ] the ability to achieve transmission speeds of 100 Mbps over UTP cable. Unlike FDDI, which supports 100 Mbps over long distances, high speed UTP technology offers 100 Mbps of transmission speed, but over limited distances (usually up to 100 meters). High speed UTP technology is the key element in the modern non-fiber 100 Mbps Ethernet and Token Ring LAN implementations.

Although 100 Mbps LANs/MANs have satisfied many of the modern requirements for high speed networking, some network applications demand even more speed, more connections, greater distances, or all three. LANs needed to evolve to address these needs and thus began to push and shove into the world of WANs.

The LAN/WAN conflict

As you can see from the discussion thus far, distributed wide area networks and centralized local area networks led separate lives of quiet desperation through most of the 1980s. In the late 1980s and early 1990s, however, the number of LANs and demand for LAN services grew dramatically. As the number of LANs increased, the requirement to interconnect multiple LANs using WAN links also increased.

Unfortunately the underlying technology for local and wide area networks is so different that moving LAN information over a WAN link is not exactly a walk through the park. Some of the key differences include the following:

- [ ] LANs usually transmit groups of bytes, called frames, from one system to another. Each frame can contain hundreds or even thousands of bytes of information. WANs, on the other hand, tend to send small amounts of information—normally several hundred bytes—at a time and rely on multiple transmissions to convey large amounts of information. Unfortunately, a solution optimized to efficiently move large blocks of information is often inefficient for moving small blocks of information.

□ WANs and LANs differ in the way that systems communicate with one another. In a LAN environment, communication is peer oriented and no single system is in control of LAN traffic. Instead, messages pass through every system in the network, and when a message reaches the system it is addressed to, that system copies the message from the network and processes it as appropriate. In contrast, systems communicating in WAN environments tend to employ a master-slave relationship, where one "master" system is in control of what messages flow to/from the "slave" systems.

□ LAN environments tend to support many more systems over a common connection than WAN environments. For example, one multi-drop WAN connection may support 16 to 24 workstations or systems, while a single LAN connection may support 100 to 200 systems. This is also why speed is so important in the LAN environment—the bandwidth to adequately service 200 systems must obviously be larger than the bandwidth required to provide the same level of service to 20 systems.

The modern rush to interconnect LANs has not, however, been slowed down by these technical differences. Instead, key technology has been developed to address the difficulties associated with moving LAN traffic over WAN links. This technology falls into three categories:

□ Network devices that optimize the transmission of LAN traffic over WAN links.
□ Special WAN services designed for interconnecting LANs.
□ High speed WAN links that offer performance superior to current LAN technology.

The first category of technology—network devices—is the traditional solution for moving LAN traffic over WAN links. These network devices include bridges, routers, and gateways. Although all of these devices enable LANs to be interconnected over WAN links, each one uses a different methodology. Bridges, for example, filter information based on the sending and receiving system addresses. Routers and gateways, on the other hand, filter information based on the network protocol in use (e.g., Novell's Internetwork Packet eXchange (IPX), IBM's System Network Architecture (SNA), Digital's DECnet, etc.).

Because bridges, routers, and gateways filter the information that flows between LANs, they can operate over "slow" analog phone lines or higher-speed digital links (including ISDN links) and still provide adequate (but not great) performance. A single physical WAN link (or dedicated ISDN connection) is required for every LAN interconnection, and compatible network devices must be placed on each end of that connection.

This type of solution works well when connecting two LANs together. When more than two LANs need to be interconnected, however, the implementation becomes far more complex because multiple WAN links and multiple network devices are required. If, for example, three LANs need to be interconnected, three WAN links (one for each unique pair of LANs) and six bridges, routers, or gateways are needed.

(2) The second category of LAN/WAN connection technology—specialized WAN services—addresses the issues of multiple LAN interconnections. Although several private services are available in this category, the best known public service is frame relay.

Conceptually, frame relay is like a packet switching network for LANs. Like the conventional packet switching WAN technology, frame relay implements a meshed network supporting multiple connections. Once a LAN is connected to the frame relay network, it can theoretically reach any other LANs attached to the frame relay network. In reality, LAN-to-LAN traffic within the network is limited to the LANs you select (and pay for). Unlike packet switching networks, frame relay networks move the entire LAN frame from one point to another.

The connection between a LAN and a frame relay network uses a device similar to a bridge or a router. In the case of frame relay, however, a single network device—and a single frame relay connection—provides interconnections for more than two LANs. This is the main attraction of frame relay networks—LANs located in multiple locations can be interconnected by installing a single, simple connection at each location.

Several vendors offer commercial frame relay services and the "insides" of the network vary from one vendor to the next. Most of them rely on a combination of satellite links and digital links to form the network itself. This means, of course, that transmission speed through the network is limited by the types of WAN links that make it up. Therefore, from an overall throughput stance, frame relay does not offer better performance than you can obtain by installing your own WAN links and network devices—it is, however, less expensive and less difficult to manage.

17

Both of the first two categories of LAN/WAN interconnection technology are dramatically affected by the speed (or more to the point, the lack of speed) available over WAN links. To put it simply, the performance and usability of a LAN-to-LAN link is directly related to the speed of that link.

The ideal solution of the LAN/WAN interconnection dilemma is to use WAN links that offer transmission speeds equal to or better than those of the LANs. If, for example, you link two 10 Mbps Ethernet LANs together over a 100 Mbps WAN link, the two LANs can truly function as a single, integrated network because no delays occur when traffic moves between them. This is the goal of the third and final category of LAN/WAN interconnection technology—high speed WAN links.

Given the state of today's electronics, the medium used to construct a high speed WAN link is of the utmost importance. For example, although copper wire can, under some circumstances, be used for speeds up to 100 Mbps, it is really impractical for long distance (wide area) applications. In fact, the best medium for high speed WAN links, that has the potential to achieve transmission speeds measured in gigabits per second (Gbps), is fiber optic cable.

The problem with a fiber optic WAN solution is how to deploy it. Clearly you cannot tear down all of the copper phone wires and replace them with fiber optic cable overnight. Furthermore, although several long distance vendors have implemented fiber optic runs, a realistic WAN solution must be well-defined and widely available. That is the intent of Broadband ISDN (B-ISDN)—the follow-up to the digital-based ISDN.

Like ISDN, B-ISDN is a single solution for both voice and computer communications, but B-ISDN users fiber optic links instead of digital links. In the first implementation of B-ISDN, speeds of 155 Mbps will be supported, and subsequent implementations are intended to raise that rate. The commercial availability of B-ISDN follows in the footsteps of ISDN, which, as noted, has yet to reach full commercial availability. The transmission technology that makes B-ISDN work, however, is available today. That technology is Asynchronous Transfer Mode (ATM).

ATM is a transmission methodology developed for the fiber optic medium. Under ATM, transmissions—voice or data—are broken down into small "cells," usually 53 bytes in length. These cells are then switched through a meshed fiber optic network similar to the way that packets move through a packet switching network and frames move through a frame relay network. The combination of the optimum cell size and the throughput of fiber optic media enables ATM to outperform other fiber-based transmission methodologies.

In fact, ATM transmission is so efficient and so popular that it has been applied to environments other than B-ISDN. Private ATM networks, for example, have been created to offer high speed services now. Even more interesting, ATM technology has been applied to the LAN environment to create networks that offer performance superior to FDDI LANs. Thus ATM is a rare case where WAN technology has moved into the LAN arena.

LAN/WAN interconnection solutions

There are no easy answers when it comes to LAN/WAN interconnections. The best solution for any particular networking problem is dictated by the network protocols you are running, the type of applications you are running, and the networking services available in your geographical area. A solution that works well to integrate two Novell Ethernet LANs in California may not be a solution for linking a TCP/IP Ethernet LAN in Kansas with a TCP/IP Ethernet LAN in North Carolina. Every situation has its own unique "spin."

Given all of these considerations, what is the best way to find the "right" answer to your LAN/WAN interconnection problems? This is an easy question, because here knowledge is the key. If you spend the time to learn about the various LAN technologies, WAN technologies, and interconnection tools—most of which are explained in this book—you will be able to make informed decisions appropriate to your unique network requirements.

19

LAN basics

AFTER YOU MAKE THE COMMITMENT TO GET INVOLVED WITH Local Area Network (LAN) technology, you are faced with the proposition of learning the differences among the various LAN technologies available. Although many people believe that Token Ring and Ethernet are the only two major LAN standards, the reality is that four major standards apply to modern LANs: Ethernet, IEEE 802.3, IBM Token Ring, and Institute of Electrical and Electronics Engineers (IEEE) 802.5.

In this group of four LAN standards, Ethernet and 802.3 are somewhat similar, and IBM Token Ring and 802.5 are nearly (but not entirely) identical. For example, systems following the Ethernet specifications can coexist on the same physical LAN as systems following the 802.3 specifications, but the two types of systems cannot communicate with one another. In contrast, systems following either the IBM Token Ring specifications or the 802.5 specifications can, in fact, communicate with one another.

Beyond the technical realm, one of the important differences between these types of LANs is philosophical in nature. Specifically, the IEEE 802.3 and 802.5 specifications are driven by the IEEE, which is a noncommercial, standards-producing organization. The IEEE reports its recommendations to the American National Standards Institute (ANSI), which contributes them in turn to the International Standards Organization (ISO).

Ethernet and IBM Token Ring, on the other hand, are commercial implementations driven, for the most part, by Digital Equipment Corporation and IBM. While this is not intrinsically bad, it is worth considering that the commercial motivation of IBM and Digital may not be as altruistic as the motivation of the IEEE. For example, for nearly a decade, IBM chose not to support or endorse Ethernet, while Digital chose not to endorse or support Token Ring. Those choices were not, however, based on the technology pros and cons of each type of LAN. Instead, they were based on how one company could position its products against the other. Love Ethernet, love Digital (and hate IBM). Love Token Ring, love IBM (and hate Digital).

Fortunately, the LAN superiority battles wore themselves out by the early 1990s. At that time Digital elected to add support for Token Ring across its product line, and IBM added support for Ethernet across its line. This trend spread beyond IBM and Digital, and now most major manufacturers support some mixture of Ethernet, IEEE 802.3, IBM Token Ring, and IEEE 802.5 specifications.

In order to appreciate the significance of having this range of choices, we need to take a look at the history of Ethernet and Token Ring LANs.

History of Ethernet /802.3

Ethernet and 802.3 are similar (but not identical) implementations of a Carrier Sense, Multiple Access with Collision Detection (CSMA/CD) LAN. Ethernet was developed in the commercial market, while 802.3 was developed by an independent standards organization. Specifically:

☐ Ethernet was created by Xerox Corporation in 1972; however, the first commercial implementation of Ethernet—Version 1— was developed under a joint effort involving Xerox, Digital Equipment Corporation, and Intel Corporation, and was introduced in 1980. The Version 1 specification was subsequently refined and introduced in 1982 as Ethernet Version 2.
☐ The 802.3 implementation was designed by the IEEE as part of its 802 series of LAN specifications. These specifications were approved by the IEEE in 1983.

The 802.3 design was, in fact, based on the Version 1 Ethernet implementation. The IEEE enhanced the design to the extent that the two implementations were incompatible with one another. In response, Xerox, Intel, and Digital Equipment created the Version 2 implementation of Ethernet so that 802.3 and Ethernet could coexist on the same physical LAN.

Because of the similarities, and the fact that both Ethernet and 802.3 can coexist on the same physical LAN (or in the same system), the two terms are often used together, or interchangeably. In point of fact they are not interchangeable, and network software using the Ethernet specifications cannot communicate with a network software using the 802.3 specifications. Additional technical information on the differences between Ethernet and 802.3 are presented in Chapter 2.

History of Token Ring/802.5

The design for Token Ring networking dates back to 1969, but Token Ring technology did not become a major factor in Local Area Networks until the mid-1980s. Specifically, in 1982 IBM presented a series of papers on Token Ring technology to the IEEE, who were beginning the development of the 802 series of LAN standards. IBM's presentation became a driving force that resulted in the development of the IEEE 802.5 standard for token-passing ring networks, released in 1985.

IBM's first foray into Token Ring networking also took place in 1985 with the announcement of a Token Ring LAN for PCs. This offering was then broadened in 1986 to include other platforms in the IBM computer family. Since that time, Token Ring has been the backbone of IBM's LAN strategy.

IBM's implementation of Token Ring and the IEEE specifications for the 802.5 standard are extremely similar, but not identical. The specific differences between IBM Token Ring and 802.5 are described in Chapter 3. To be sure, an IBM Token Ring network can be installed and operated according to the 802.5 specifications—IBM has simply chosen to "tweak" the specification in certain areas. In general, the differences between IBM's implementation and the IEEE specifications are not worth losing any sleep over.

IEEE 802.2

The commercial implementations of Ethernet and Token Ring have very little in common. The IEEE implementations of 802.3 and 802.5, however, have a great deal in common because the IEEE wanted to create one framework that would embrace multiple LAN implementations. To accomplish this, the IEEE broke LAN functions into two general areas:

23

☐ Logical Link Control (LLC)—This interface defines how the LAN interacts with higher-level network protocols (e.g., SNA, IPX).

☐ Media Access Control (MAC)—This interface relates to the LAN access methodology (token-passing or carrier-sensing) and the physical aspects of the LAN.

Under the IEEE strategy, the LLC interface is common to all LAN implementations—an IEEE 802.3 CSMA/CD LAN uses the same LLC structure as an IEEE 802.5 token-passing ring LAN. Under this same strategy, the MAC interface is different for each LAN implementation. This architecture is shown in Fig. II-1.

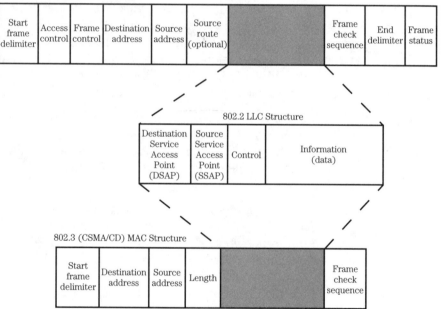

■ II-1 *Common LLC/different MACs*

The structure of the LLC interface is described by the IEEE 802.2 specifications. Each MAC-level interface is described by its own specifications, with the two most popular being the IEEE 802.3 and 802.5 specifications for CSMA/CD and token-passing LANs, respectively. Other MAC-level members of the 802 family include the 802.4 specifications for a token-passing bus and 802.6 specifications for a fiber-based metropolitan area network.

802.2 LLC Structure

Destination Service Access Point (DSAP) [8 bits]	Source Service Access Point (SSAP) [8 bits]	Control [8-16 bits]	Information (data) [variable length]

■ **II-2** *802.2 Field layout*

As shown in Fig. II-2, the 802.2 structure contains information that allows network-level applications to quickly identify the contents of the MAC-level structure. The following individual components make up the 802.2 structure:

☐ Destination Service Access Point (DSAP)—This one-byte field identifies the network protocol that should be used by the receiving system to interpret the message. See the discussion that follows these 802.2 field definitions for more information on the DSAP.

☐ Source Service Access Point (SSAP)—This one-byte field identifies the network protocol used to originate the message. Normally the SSAP contains the same value as the DSAP. See the discussion that follows these 802.2 field definitions for more information on the SSAP.

☐ Control—This field is used for various commands, including exchange identifier, test, connect, disconnect, and frame reject. The Control field can be one or two bytes long, with the length defined by the first two bits.

☐ Data—In one sense, this field carries the user data that is moving through the network. In a broader sense, however, this field contains the structures and information required by the higher-level network protocol(s) being used in the network. For example, in a Novell NetWare network, the data field will carry the IPX information, including any actual user data.

As noted above, the DSAP and SSAP fields serve to identify the originating (source) network protocol (SSAP) and the target (destination) network protocol (DSAP). In the vast majority of cases, the SSAP and the DSAP contain the same SAP value. For example, SAP assignments designate the following network protocols:

SAP	Assignment
04-05	IBM SNA
06	ARPANET Internet Protocol (IP)
18	Texas Instruments
7E	X.25 Level 3
80	Xerox Networking Services (XNS)
98	Address Resolution Protocol (ARP)
BC	Banyan VINES
E0	Novell NetWare
F0	IBM NetBIOS
F4-F5	IBM LAN management
F8	IBM Remote Program Load (RPL)
FA	Ungermann-Bass
FE	Open System Interconnect (OSI) network layer

When a system receives a message, it looks at the DSAP value to make sure that it supports the network protocol in use. This is particularly important for a system that supports multiple network protocols, because the value of the DSAP will determine the correct software module to handle the message. Similarly, if a system receives a message that uses a network protocol it does not support, the system can ignore the message.

As you can see from this discussion, the SSAP and DSAP are key components in implementing multiple network protocols in a single physical LAN. This topic is further discussed in Chapter 4.

Choosing a LAN implementation

Most market vendors no longer impose artificial constraints on physical LAN selection. You can, for example, run IBM SNA, Digital DECnet, Novell IPX, and TCP/IP in Ethernet, IEEE 802.3, IEEE 802.5, or IBM Token Ring environments. Unfortunately, this freedom of choice creates a slightly different problem: choosing the right LAN for the set of network protocols and network applications you have selected.

Of course there is no single, simple solution to that problem. Every environment has its own special constraints and considerations. A better way to look at the issue is to explore the questions commonly asked during the LAN selection process. In particular, the following questions serve to highlight the strengths and weaknesses of Ethernet/802.3 and IBM Token Ring/802.5:

Which is easier to troubleshoot?

Token Ring. The Token Ring access protocol includes test procedures that are invoked every time a system first accesses the ring. When one of the procedures fails, messages are reported indicating probable causes of failure, including suspicious addresses. Ethernet has no such provisions, and is, for the most part, diagnosed through a combination of isolation procedures and will power.

Which is easier to install?

Ethernet. The flexibility to mix and match hub (or star), bus, and tree topologies makes it extremely easy to plan and install an Ethernet network of any size. This is especially true with thin-cable Ethernet, where systems can be daisy-chained together with T-connectors. In contrast, Token Ring forces you to use a hub topology where you must plan connections back to the main Multistation Access Unit (MAU) for every system or every hub.

Which is best for IBM connections?

Token Ring at 16 Mbps. A 16 Mbps Token Ring network using 16 kB frames will outperform an Ethernet network, which uses 1518-byte frames running at 10 Mbps. Of course, you need to be carrying out operations that will actually fill the frame to realize any benefits. If, for example, you are doing character-oriented terminal access (for example, using a VT100 emulation package), you won't notice much of a difference between Ethernet and Token Ring. But if you're doing file transfer using SNA services, you should see a dramatic difference.

Which is best for a multi-vendor environment?

Ethernet. Most non-IBM systems support Ethernet as the primary LAN of choice. For example, Digital Equipment, Hewlett-Packard, and Sun Microsystems networks are normally implemented using Ethernet. Even IBM's RS/6000 system comes with Ethernet—Token Ring is a priced option. So although Token Ring is "officially" supported by most companies, Ethernet remains the mainstream preference.

Which supports more stations?

Debatable. Anyone who implements a self-contained network with more than 200 systems is pushing the limits of reality. A better strategy is to implement small segments/rings and interconnect

27

them as needed. In this regard, both Ethernet and Token Ring can be easily carved up into small LANs and then interconnected.

Which offers better performance?

Go fish. Ethernet is better suited for LAN environments where LAN activities are random and "bursty" in nature. Token Ring favors environments where access is methodical and predictable. In theory, a 10 Mbps Ethernet network will outperform a 4 Mbps Token Ring network, and a 16 Mbps Token Ring network will outperform a 10 Mbps Ethernet network. But in reality, the actual performance is mitigated by the average amount of data being transmitted, the number and locations of devices transmitting, and the frequency at which each device transmits. In other words, the raw throughput of a LAN is only one factor in determining overall LAN performance.

Which has a better future?

Tie. The sheer number of installed Ethernet and Token Ring networks guarantees that both types of LANs will be significant factors in the ongoing evolution of LAN technology. For example, Token Ring has been a driving force in the evolution of 100 Mbps Fiber Distributed Data Interface (FDDI), and Ethernet has been the motivation behind development of 100 Mbps 100BASE-VG. Of course both of these 100 Mbps solutions will require new adapters, hubs, repeaters, and cabling, but the spirit of Token Ring and Ethernet will live on within them. Just remember—nothing lasts forever.

Finally, please note that this discussion has provided only a glimpse of the technical composition of the Ethernet, IEEE 802.3, IBM Token Ring, and IEEE 802.5 LAN specifications. More detailed information is presented in the next two chapters.

28

Ethernet/802.3

2

Access discipline

As described in the introduction to Part II, Ethernet and IEEE 802.3 are similar in both nature and function. For example, both Ethernet and 802.3 specify LAN speeds of 10 Mbps and use frames that can carry up to 1500 bytes of user data.

The most significant similarity, however, is that both types of LANs rely on a "carrier sensing" mechanism to allow multiple systems to contend for LAN access, and on a "collision detection" methodology to enable systems to detect (and recover) from data collisions. These two mechanisms—carrier sensing and collision detection—form the core technology of the Carrier Sensing, Multiple Access with Collision Detection (CSMA/CD) access discipline.

Under CSMA/CD, a system that wants to make a transmission "listens" to the LAN to see if other systems are currently transmitting data. The system listens by testing the electrical level of the receive circuit for the presence of a carrier signal—this is the "carrier sensing" aspect of CSMA/CD.

If the LAN is quiescent (no electrical activity is detected on the receive circuit), the system can transmit a frame of information. On the other hand, if the LAN is busy, the system must wait for a brief period of time and then retest the condition of the LAN. This testing process may be repeated many times before the system is able to transmit its information, but fortunately the tests are performed on a microsecond time scale.

In the event that two systems detect an idle LAN at the same time and start transmitting at the same time, a data collision occurs and the information is garbled beyond recognition. The transmitting systems recognize that a collision has occurred because of the electrical disturbance generated by both systems transmitting at the same time—the signal level is twice what it should be.

After experiencing a data collision, each system waits for a small random interval of time before attempting to transmit again. If another collision occurs on the next attempt, the system will wait for a longer period of time. This process is repeated until the information is successfully transmitted or the operation is aborted.

As you can see, data collisions can only occur when two or more systems begin transmitting at the same time, because if one system accesses the LAN first, others will detect a carrier and will not attempt to transmit. Once a data frame is successfully transmitted (no collisions are reported), Ethernet regards the network transaction as complete. No acknowledgments are generated at the CSMA/CD protocol level to indicate if the frame was successfully received or not.

In many ways the CSMA/CD access discipline is analogous to human conversation. Think of the way people converse in a group. If people want to talk, they first make sure no one else is talking. If the airways are silent, someone goes ahead and speaks. If two people start talking at the same time, they hear the "collision" and stop. After a few seconds—and possibly after one or two more false starts—one of them gets control of the conversation.

Now think of how the CSMA/CD access discipline follows this pattern. Systems compete for access to the LAN medium when they have something to transmit; otherwise they listen for incoming messages. If only one or two systems have information to send, they can dominate the LAN until they are done (or until other systems have information to send). Conversely, if many systems have information to transmit, they must all politely vie for the LAN, waiting, in effect, for their turn to "talk."

The casual, uneven nature of CSMA/CD access is both a positive point and a negative point, depending on traffic. On the positive side, CSMA/CD access allows a minority of devices to dominate a LAN if other systems have minimal or no information to transmit. For example, if only two systems in a 100-system LAN have information to exchange, they can rule the LAN for the duration of the exchange.

On the negative side, CSMA/CD access can produce periods of unusually slow response times in a heavily loaded LAN. When LAN activity is high, systems must wait for LAN access (wait for other systems to stop transmitting) and they also experience a high number of data collisions. Both of these factors have a negative effect on LAN performance. For example, imagine 50 out of 100 systems trying to transmit within a few microseconds of each other—each system must compete with 49 other stations for access.

Addressing scheme

Once a system gets access to the physical LAN, it can then transmit a frame of information over the LAN. Although the specific construction of a LAN frame differs between Ethernet and 802.3, as shown later in this chapter, both frame formats share these logical elements:

☐ A source address that identifies the system originating the frame.
☐ A destination address that identifies the system (or group of systems) the frame is intended for.
☐ An indication of what network protocol (e.g., TCP/IP, IPX, SNA, DECnet) is carried in the frame.
☐ A data area that contains the network protocol information along with any "real" data.
☐ A checksum that allows the receiver(s) to verify that the frame has arrived intact.

The source and destination addresses correspond to low-level (MAC-level) hardware addresses of the network interface card (or addresses overlaid on top of the hardware addresses). These addresses are the "bottom line" for getting a frame to a system. For more information on how these hardware addresses relate to higher-level network addresses, please consult Chapter 4.

Ethernet and 802.3 use a 48-bit address scheme normally represented as six pairs of hexadecimal values. For example, 08-00-2B-05-1C-62, AA-00-04-00-02-A0, and 00-00-C0-23-4E-29 are all valid Ethernet/802.3 addresses. Each system must have a unique address to participate in an Ethernet/802.3 network. That address is normally assigned by the hardware manufacturer and programmed directly into the network adapter hardware, but many software drivers allow that address to be overridden with a "soft" address.

The first three pairs of an Ethernet/802.3 address determine the identity of the manufacturer. The IEEE assigns each manufacturer a three-pair value, or a range of values, that they use as a prefix for all of their network adapters. This assignment brings a useful structure to the LAN, because you can determine what type of equipment is originating or receiving a frame just by looking at the address (unless, of course, that address has been overridden).

31

In IEEE terminology, the assigned address is referred to as an Organizationally Unique Identifier (OUI), and is applicable to any IEEE-approved LAN or Metropolitan Area Network (MAN). Over a thousand OUIs have been assigned, but only about half of those assignments are available as public information. A partial list of these IEEE OUIs is shown in Table 2-1.

■ **Table 2-1 IEEE address prefixes.**

OUI	Company
00-00-0E	FUJITSU LIMITED
00-00-1B	NOVELL, INC.
00-00-3F	SYNTREX, INC.
00-00-46	OLIVETTI NORTH AMERICA
00-00-4C	NEC CORPORATION
00-00-52	OPTICAL DATA SYSTEMS
00-00-58	RACORE COMPUTER PRODUCTS, INC.
00-00-61	GATEWAY COMMUNICATIONS
00-00-65	NETWORK GENERAL CORPORATION
00-00-6D	CRAY COMMUNICATIONS, LTD.
00-00-6E	ARTISOFT, INC.
00-00-6F	MADGE NETWORKS, LTD.
00-00-74	RICOH COMPANY, LTD.
00-00-79	NETWORTH, INC.
00-00-7D	CRAY RESEARCH SUPERSERVERS, INC.
00-00-7F	LINOTYPE-HELL AG
00-00-94	ASANTE TECHNOLOGIES
00-00-98	CROSSCOMM CORPORATION
00-00-99	MEMOREX TELEX CORPORATION
00-00-A7	NETWORK COMPUTING DEVICES, INC.
00-00-A8	STRATUS COMPUTER, INC.
00-00-A9	NETWORK SYSTEMS CORP.
00-00-AA	XEROX CORPORATION
00-00-BC	ALLEN-BRADLEY CO., INC.
00-00-C0	SMC (WESTERN DIGITAL)
00-00-D0	DEVELCON ELECTRONICS, LTD.
00-00-E7	STAR GATE TECHNOLOGIES
00-00-E8	ACCTON TECHNOLOGY CORP.
00-00-F2	SPIDER COMMUNICATIONS
00-00-F8	DIGITAL EQUIPMENT CORPORATION
00-07-01	RACAL-DATACOM
00-20-09	PACKARD BELL ELECTRONICS, INC.
00-20-1D	KATANA PRODUCTS
00-20-24	PACIFIC COMMUNICATION SCIENCES
00-20-33	SYNAPSE TECHNOLOGIES, INC.
00-20-35	CPN ARCHITECTURES/IBM CORP.
00-20-48	FORE SYSTEMS, INC.

OUI	Company
00-20-4E	NETWORK SECURITY SYSTEMS, INC.
00-20-51	PHOENIX DATA COMMUNICATIONS
00-20-58	ALLIED SIGNAL, INC.
00-20-5C	INTERNET SYSTEMS/FLORIDA, INC.
00-20-61	DYNATECH COMMUNICATIONS, INC.
00-20-64	PROTEC MICROSYSTEMS, INC.
00-20-66	GENERAL MAGIC, INC.
00-20-69	ISDN SYSTEMS CORPORATION
00-20-75	MOTOROLA COMMUNICATION ISRAEL
00-20-7A	WISE COMMUNICATIONS, INC.
00-20-80	SYNERGY (UK) LTD.
00-20-94	CUBIX CORPORATION
00-20-9A	THE 3DO COMPANY
00-20-AF	3COM CORPORATION
00-20-B0	GATEWAY DEVICES, INC.
00-20-BE	LAN ACCESS CORP.
00-20-D2	RAD DATA COMMUNICATIONS, LTD.
00-20-D9	PANASONIC TECHNOLOGIES, INC.
00-20-EE	GTECH CORPORATION
00-20-FB	OCTEL COMMUNICATIONS CORP.
00-40-0D	LANNET DATA COMMUNICATIONS, LTD.
00-40-27	SMC MASSACHUSETTS, INC.
00-40-32	DIGITAL COMMUNICATIONS
00-40-4F	SPACE & NAVAL WARFARE SYSTEMS
00-40-66	HITACHI CABLE, LTD.
00-40-6F	SYNC RESEARCH, INC.
00-40-76	AMP INCORPORATED
00-40-9D	DIGIBOARD, INC.
00-40-9F	LANCAST/CASAT TECHNOLOGY, INC.
00-40-A6	CRAY RESEARCH, INC.
00-40-B4	3COM K.K.
00-40-DF	DIGILOG SYSTEMS, INC.
00-40-FF	TELEBIT CORPORATION
00-60-8C	3COM CORPORATION
00-80-00	MULTITECH SYSTEMS, INC.
00-80-06	COMPUADD CORPORATION
00-80-15	SEIKO SYSTEMS, INC.
00-80-19	DAYNA COMMUNICATIONS, INC.
00-80-1A	BELL ATLANTIC
00-80-21	NEWBRIDGE RESEARCH CORP.
00-80-23	INTEGRATED BUSINESS NETWORKS
00-80-24	KALPANA, INC.
00-80-26	NETWORK PRODUCTS CORPORATION
00-80-3E	SYNERNETICS
00-80-45	MATSUSHITA ELECTRIC IND. CO.
00-80-51	FIBERMUX
00-80-74	FISHER CONTROLS

OUI	Company
00-80-7B	ARTEL COMMUNICATIONS CORP.
00-80-8A	SUMMIT MICROSYSTEMS CORP.
00-80-90	MICROTEK INTERNATIONAL, INC.
00-80-A3	LANTRONIX
00-80-AD	CNET TECHNOLOGY, INC.
00-80-AE	HUGHES NETWORK SYSTEMS
00-80-C2	IEEE 802 COMMITTEE
00-C0-00	LANOPTICS, LTD.
00-C0-0D	ADVANCED LOGIC RESEARCH, INC.
00-C0-13	NETRIX
00-C0-1C	INTERLINK COMMUNICATIONS, LTD.
00-C0-1D	GRAND JUNCTION NETWORKS, INC.
00-C0-33	TELEBIT COMMUNICATIONS APS
00-C0-34	DALE COMPUTER CORPORATION
00-C0-43	STRATACOM
00-C0-63	MORNING STAR TECHNOLOGIES, INC.
00-C0-64	GENERAL DATACOMM IND., INC.
00-C0-6D	BOCA RESEARCH, INC.
00-C0-6F	KOMATSU, LTD.
00-C0-77	DAEWOO TELECOM, LTD.
00-C0-A8	GVC CORPORATION
00-C0-AA	SILICON VALLEY COMPUTER
00-C0-B0	GCC TECHNOLOGIES, INC.
00-C0-B3	NORAND CORPORATION
00-C0-B6	MERIDIAN DATA, INC.
00-C0-E2	CALCOMP, INC.
00-C0-E8	PLEXCOM, INC.
00-C0-EC	DAUPHIN TECHNOLOGY
00-C0-ED	US ARMY ELECTRONIC
00-C0-EE	KYOCERA CORPORATION
00-C0-F3	NETWORK COMMUNICATIONS CORP.
00-C0-F4	INTERLINK SYSTEM CO., LTD.
00-C0-FB	ADVANCED TECHNOLOGY LABS
02-07-01	RACAL-DATACOM
08-00-07	APPLE COMPUTER, INC.
08-00-09	HEWLETT PACKARD
08-00-0B	UNISYS CORPORATION
08-00-1F	SHARP CORPORATION
08-00-28	TEXAS INSTRUMENTS
08-00-2B	DIGITAL EQUIPMENT CORPORATION
08-00-39	SPIDER SYSTEMS LIMITED
08-00-5A	IBM CORPORATION
08-00-69	SILICON GRAPHICS, INC.
08-00-74	CASIO COMPUTER CO. LTD.
08-00-8F	CHIPCOM CORPORATION

34

OUI	Company
10-00-5A	IBM CORPORATION
AA-00-00	DIGITAL EQUIPMENT CORPORATION
AA-00-03	DIGITAL EQUIPMENT CORPORATION
AA-00-04	DIGITAL EQUIPMENT CORPORATION

This list of Organizationally Unique Identifiers is updated quarterly. A current list of public OUIs is available from the Institute of Electrical and Electronics Engineers (IEEE).

The remaining three pairs in an Ethernet/802.3 address are then assigned by each manufacturer to create a unique address for that specific adapter. The resulting six-pair adapter address forms the unique address for a specific system. That address is used as the source address for all traffic originated by that system, and as the destination address for traffic directed to that specific system.

In addition to specific system addresses, Ethernet and 802.3 also support two types of group addressing. The first, and easiest to understand, is a "broadcast" address. A broadcast address is composed of binary ones (i.e., FF-FF-FF-FF-FF-FF) and every system that receives a broadcast address attempts to process it.

The second type of address is a "multi-cast" address. Multi-cast addresses are group addresses that are received by a series of like-kind machines. A multi-cast address is constructed by adding a "1" to the first pair of hexadecimal values in the six-pair address. For example, the multi-cast address for a Cisco device would begin 01-00-0C. In reality, multi-cast addresses are not commonly used.

Topology & construction

As shown by Fig. 2-1, Ethernet and 802.3 networks can be configured using a bus topology, tree topology, hub (or star) topology, or any combination of the three. Large Ethernet and 802.3 networks normally begin with a main bus cable that establishes the "backbone" of the LAN, and then employ tree and/or hub segments originating at the main backbone to reach groups of systems.

Connecting a device to an Ethernet LAN requires the use of an attachment device called a transceiver. A transceiver can be an external device or transceiver functions can be integrated into a network adapter. As a general rule, this is a function of the type of medium—connection to the thick coaxial backbone is typically handled by an external transceiver, while connections to thin coaxial or twisted-pair segments can be internal or external.

35

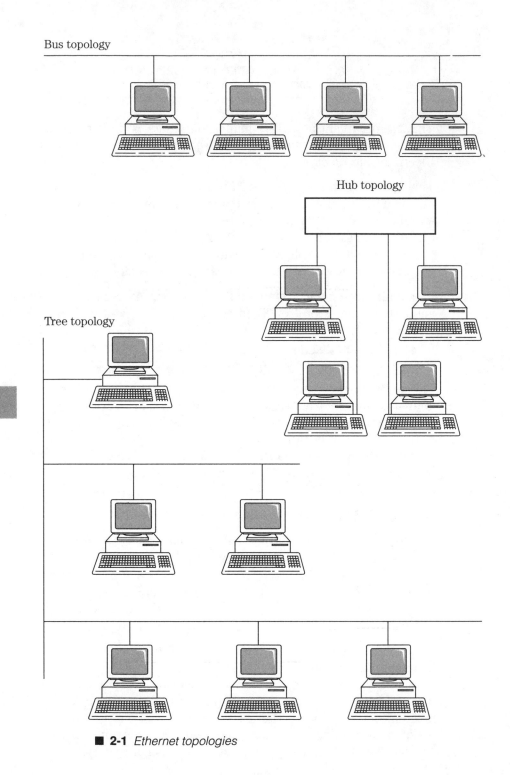

Bus topology

Hub topology

Tree topology

■ **2-1** *Ethernet topologies*

The term "transceiver" (TRANSmitter/reCEIVER) is for the most part an Ethernet term. In the formal specifications for the IEEE 802.3 network, the attachment functions are handled by a Medium Attachment Unit (MAU). The popularity of the term transceiver has, however, resulted in a spillover of its use in the 802.3 environment. The primary function of the transceiver/MAU is to translate the digital information generated by the system into the electrical format appropriate to the connection medium. Beyond managing send and receive operations, a transceiver (or the transceiver logic on an interface adapter) performs several functions of particular interest and value to Ethernet and 802.3 LANs:

☐ Collision detection: The transceiver contains the circuitry that detects when a collision occurs on the medium. When a collision is detected, the transceiver performs two actions. First, the transceiver generates a signal to notify the transmitting system that a collision has occurred. Second, after notifying the transmitter about the collision, the transceiver then broadcasts a "jamming" signal on the LAN. The jamming signal alerts other systems that a collision has occurred and also allows the LAN to "settle" before transmission resumes.

☐ Heartbeat: A heartbeat is a brief signal generated by the transceiver after every transmission to ensure the main adapter that the transmission was free of collisions. A heartbeat uses the same electrical interface used to report collisions, but notification occurs *after* transmission and is much shorter in duration. Although the heartbeat function is defined in both Ethernet (as the Collision Presence Test) and 802.3 specifications (as the Signal Quality Error (SQE) Test), it is rarely used because many adapters confuse a heartbeat with a collision signal.

☐ Jabber: The jabber function allows the transceiver to discontinue transmission if the frame being transmitted exceeds the specified limit (1518 bytes). This prevents a malfunctioning system or adapter from flooding the LAN with inappropriate data.

☐ Monitor: The final function allows the transceiver to prohibit transmit functions while enabling receive and collision detection functions. As a result, the attached system can monitor LAN traffic, but it cannot originate any traffic.

Smaller LANs are often constructed solely with cable, connectors, terminators, and transceivers. Large networks, however, usually employ some or all of the following devices:

☐ Repeaters—Repeaters are used to join two segments in order to form a single logical segment. Repeaters do not filter the traffic between segments (which can, by the way, cause congestion and complicate problem determination). Multiple repeaters can be used to interconnect consecutive segments.

☐ Multiport repeaters—A multiport repeater is a hub that allows the connection of systems or additional LAN segments to a central point. A single multiport repeater can be used to create a small stand-alone network using a hub topology.

☐ Local bridges—Given that all systems compete for LAN access under the CSMA/CD access discipline, Ethernet and 802.3 networks are often broken down into smaller network segments that are linked together using intelligent local bridges. When bridges are employed, local traffic on a LAN segment is isolated from the other LAN segments—only traffic destined for the other LAN segment moves across the bridge. This dramatically reduces network-wide competition for access to the main LAN, because systems compete for access to their local segment, and the bridge contends for access to another LAN if needed.

☐ Wide area bridges—Wide area bridges perform the same function as local bridges, but they connect the two LANs (or LAN segments) over wide area links such as X.25 packet switching networks, leased analog or digital phone lines, frame relay networks, and so on.

☐ Routers—Routers are similar to wide area bridges because they too interconnect multiple Ethernet networks over wide area links. The level at which bridges and routers operate is, however, quite different. Specifically, a bridge is oblivious to the network protocol in use (i.e., TCP/IP, IPX, DECnet, etc.), and makes its decisions on when to bridge and when not to bridge based on the Ethernet/802.3 frame information alone. In contrast, a router interacts at the network protocol level and forwards frames based on the network protocol type or information contained in the network protocol information, such as logical addresses.

Please see Chapter 4 for additional information on bridges and routers.

Additional devices are available to convert media types, to extend Ethernet links using fiber optic cable, and to perform other specialized functions.

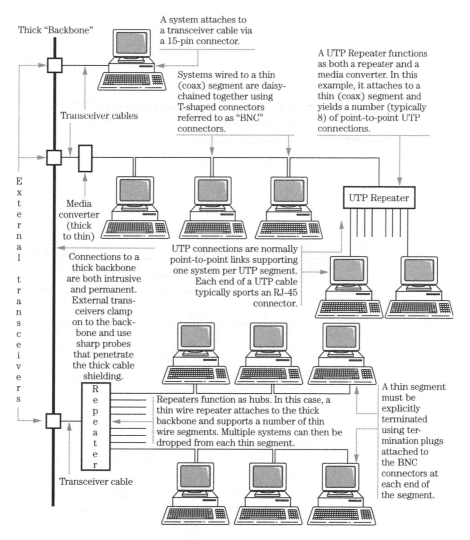

Thick "Backbone"

A system attaches to a transceiver cable via a 15-pin connector.

A UTP Repeater functions as both a repeater and a media converter. In this example, it attaches to a thin (coax) segment and yields a number (typically 8) of point-to-point UTP connections.

Systems wired to a thin (coax) segment are daisy-chained together using T-shaped connectors referred to as "BNC" connectors.

Transceiver cables

External transceivers

Media converter (thick to thin)

Connections to a thick backbone are both intrusive and permanent. External trans-ceivers clamp on to the back-bone and use sharp probes that penetrate the thick cable shielding.

UTP Repeater

UTP connections are normally point-to-point links supporting one system per UTP segment. Each end of a UTP cable typically sports an RJ-45 connector.

Repeater

Repeaters function as hubs. In this case, a thin wire repeater attaches to the thick backbone and supports a number of thin wire segments. Multiple systems can then be dropped from each thin segment.

A thin segment must be explicitly terminated using ter-mination plugs attached to the BNC connectors at each end of the segment.

Transceiver cable

■ **2-2** *Ethernet media types*

Cabling

Physical construction of an Ethernet or 802.3 LAN is as much de-pendent on the type of medium as it is on the overall topology. Fig-ure 2-2 shows the most commonly used types of media:

☐ Thick cable (10BASE-5)—As the name implies, this is a thick (10 millimeters in diameter) coaxial cable that can be up to 500 meters long per segment. Thick cables are usually employed as a main LAN backbone. External transceivers are then used to make the connection between the main LAN cable and the attached systems or devices. Each transceiver (sometimes also referred to as a "vampire tap") is permanently attached to the thick cable and presents a 15-pin D-shell connector for attachment to a system, repeater, or other device.

☐ Thin cable (10BASE-2)—Thin cable is common coaxial cable (e.g. RG-58A/U) that can be up to 185 meters long per segment. Thin cable uses twist-on BNC connectors to attach to a series of systems or devices. Each system or device attachment uses a T-shaped BNC connector to daisy-chain the connection to the next system or device. The final "T" at both ends of the chain must be fitted with a twist-on termination plug. In most 10BASE-2 implementations, transceiver functions are integrated into the LAN adapter card.

☐ Unshielded Twisted Pair (10BASE-T)—UTP cable is very similar to twisted pair phone cable. A UTP segment can be up to 100 meters long and attaches to a single system or device. Because of this single attachment design, UTP connections are typically used in conjunction with a multiport repeater (sometimes called a "hub" in this context) that provides a number of UTP connections. The physical connection is normally made with an 8-wire RJ-45 modular connector. UTP implementations can use either internal or external transceivers, depending on the systems and devices being interconnected.

Ethernet and 802.3 both use the Manchester encoding scheme to transmit and receive data on the LAN. In thick and thin cable, these signals are carried over the coaxial media. When transceiver cables or UTP cables are employed, these signals are carried over copper wire pairs. For example, a UTP cable is composed of two sets of twisted-pair wires. The assignment for these wires using a standard RJ-45 connector is as follows:

■ **Typical 10BASE-T cable.**

Wire	Assignment
1	Transmit +
2	Transmit −
3	Receive +
4	
5	
6	Receive −
7	
8	

As shown in Fig. 2-3, wires 1 and 2 should be twisted together, as should wires 3 and 6. This is a subset of the four-pair 10BASE-T implementation where 1 and 2, 3 and 6, 4 and 5, and 7 and 8 are twisted together in pairs.

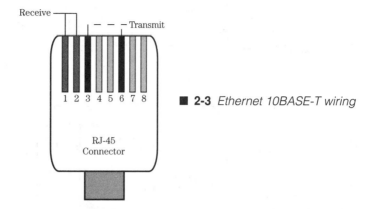

■ **2-3** *Ethernet 10BASE-T wiring*

In contrast, a transceiver cable uses a 15-pin connection. The assignment of these pins varies slightly between Ethernet and 802.3, as will be shown in the next section.

These different styles of media can be interconnected using repeaters. For example, one or more thin wire segments can be connected to a thick wire segment. A UTP multiport repeater can be attached to a thick or thin segment. A UTP segment can even be converted into a thin wire segment. This flexibility makes installing, expanding, or reconfiguring an Ethernet or 802.3 network a relatively easy and painless task.

Ethernet vs. 802.3

Both Ethernet and 802.3 specifications can coexist in a common LAN environment, but they are not completely compatible. The two specifications differ in two key areas, namely frame format and transceiver cable circuit assignments.

In terms of frame format, Ethernet and 802.3 are similar but not identical. Although they are similar enough to coexist in the same LAN, network software must explicitly interpret either the Ethernet frame or the 802.3 frame, or it must have sufficient digital intelligence to handle both formats. Figure 2-4 shows the general

41

IEEE 802.3 (CSMA/CD) frame

Preamble [7 bytes]	Start frame delimiter [1 byte]	Destination address [6 bytes]	Source address [6 bytes]	Length [2 bytes]	Destination Service Access Point (DSAP) [1 byte]	Source Service Access Point (SSAP) [1 byte]	Control [1-2 bytes]	Information (data) and optional pad [variable length]	Frame check sequence [4 bytes]

Ethernet frame

Preamble [7 bytes]	Start frame delimiter [1 byte]	Destination address [6 bytes]	Source address [6 bytes]	Type [2 bytes]	Information (data) and optional pad [variable length]	Frame check sequence [4 bytes]

■ **2-4** *Ethernet versus 802.3*

layout; the specific breakdown of the frame fields for 802.3 and Ethernet is as follows:

802.3/802.2 fields

Preamble—A seven-byte pattern of binary 1s and 0s to establish synchronization.

Start Frame Delimiter—A one-byte pattern indicating the start of the frame.

Destination Address—The address of a specific station or group of stations to receive the frame. The formal specifications allow for addresses to be two or six bytes long, but most implementations support an address size of six bytes (48 bits).

Source Address—The address of the sending station. Can be two or six bytes long. Again, most implementations use an address size of six bytes.

Length—A two-byte field containing the length of the 802.2 structure (header and data).

DSAP—The one-byte 802.2 Destination Service Access Point. See the discussion under the heading *IEEE 802.2* in the introductory section for Part II.

SSAP—The one-byte 802.2 Source Service Access Point. See the discussion under the heading *IEEE 802.2* in the introductory section for Part II.

Control—An 802.2 field used for various commands, including exchange identifier, test, connect, disconnect, and frame reject. Can be one or two bytes long, with the length defined by the first two bits.

Data—The actual information being transmitted.

Pad Bytes—Optional nondata bytes added to ensure that the frame meets minimum length requirements.

Frame Check Sequence—A four-byte field containing a checksum of the frame information beginning with the destination address and ending with the data.

Ethernet fields

Preamble—Same as corresponding 802.3 field.

Start Frame Delimiter—Same as corresponding 802.3 field.

Destination Address—The six-byte address of a specific station or group of stations to receive the frame.

Source Address—The six-byte address of the sending station.

Type—A two-byte field that identifies the network protocol or protocol service carried in the frame.

Data—The actual information being transmitted.

Frame Check Sequence—Same as corresponding 802.3 field.

One of the key differences between the two frame formats is that Ethernet relies on a single "Type" field to identify the type of network or network protocol it is carrying. In contrast, the 802.3 frame uses the 802.2-level "Source Service Access Point" (SSAP) and "Destination Service Access Point" (DSAP) to determine what network the frame came from and what network it is going to. For more information on SAPs, please refer to the discussion under the heading *IEEE 802.2* in the introductory section for Part II.

Well over 100 Ethernet "Types" are currently registered, and a partial list of these assignments is shown in Table 2-2.

■ Table 2-2 Ethernet "type" assignments.

Type	Assignment
0600	Xerox Network Services (XNS)
0800	ARPANET Internet Protocol (IP)
0801	X.75 Internet Protocol (IP)
0802	NBS Internet Protocol (IP)
0803	ECMA Internet Protocol (IP)
0804	CHAOSnet
0805	X.25 Level 3
0806	Address Resolution Protocol (ARP)
0807	Xerox XNS network protocol
0BAD	Banyan VINES
6000-6009	Digital Equipment Corporation
6010-6014	3Com Corporation
7000-7002	Ungermann–Bass
7030	Proteon
7034	Cabletron
8005	Hewlett-Packard probe protocol
8010	Exelan
8013-8016	Silicon Graphics
8035	Reverse ARP
8038-8042	Digital Equipment Corporation
8046-8047	AT&T
807D-8080	Vitalink
8088-808A	Xyplex
809B	Ethertalk
80C0-80C3	Digital Communication Associates (DCA)
80D5	IBM SNA
80F2	Retix
80F3-80F5	Kinetics
80F7	Apollo Computer (HP)
80FF-8103	Wellfleet Communications
8137-8138	Novell

Both Ethernet and 802.3 networks use the same physical 15-pin transceiver connector shown in Fig. 2-5, but Ethernet and 802.3 differ with respect to circuit assignments:

■ Transceiver cable.

Pin	802.3	Ethernet
1	Control In Ground	Ground
2	Control In A	Collision Presence +
3	Data Out A	Transmit +
4	Data In Ground	
5	Data In A	Receive +
6	Voltage Common	

7	Control Out A	
8	Control Out Ground	
9	Control In B	Collision Presence –
10	Data Out B	Transmit –
11	Data Out Ground	
12	Data In B	Receive –
13	Power	
14	Power Ground	
15	Control Out B	

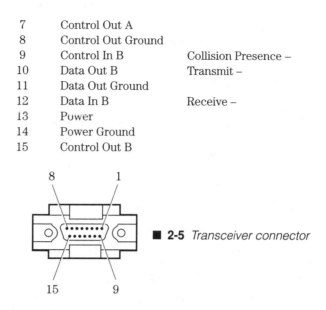

■ **2-5** *Transceiver connector*

The 802.3 specification defines four circuits, with each circuit composed of three wires (A, B, and ground). Of these circuits, one is used to transmit (Data Out), one to receive (Data In), and one for collision detection (Control In). The fourth circuit (Control Out) is currently unused. In contrast, Ethernet defines three circuits (transmit, receive, and collision), which share a common ground.

The circuits common to Ethernet and 802.3 share the same pin assignments. Although this means that the same cable can function in either environment, using one type of cable in the other environment is not recommended due to the differences in the grounding methods.

Overall specifications

General

Speed	10 Mbps*
Maximum frame size	1518 bytes
Maximum data unit	1460 bytes (802.3), 1500 bytes (Ethernet)

* IEEE 802.3 was designed to accommodate operation at 20 Mbps, but no commercial implementation of that rate is available.

Thick coax medium

Specification	10BASE-5
Impedance	50 ohms
Connector (transceiver)	DB-15
Maximum segment length	500 meters

Maximum attachments per segment	100
Minimum distance between nodes	2.5 meters
Combined length of repeated segments	2500 meters

Thin coax medium

Specification	10BASE-2
Impedance	50 ohms
Connector	BNC
Maximum segment length	185 meters
Maximum attachments per segment	30
Minimum distance between nodes	0.5 meters
Combined length of repeated segments	925 meters

UTP medium

Specification:	10BASE-T
Impedance:	85–110 ohms
Connector:	RJ-45
Maximum segment length	100 meters
Maximum nodes per segment	1
Minimum distance between nodes	not applicable
Combined length of repeated segments	2500 meters

High-speed Ethernet/802.3 alternatives

In many LANs, Ethernet/802.3 performance has reached its upper limit. Rather than moving the network to a new medium (such as fiber optic) or new topology (e.g., FDDI), three different technologies are available to bolster performance. Although these technologies will be discussed in Ethernet terms, they are also applicable to 802.3 networks.

In general terms, these performance-improving approaches are:

☐ Ethernet switching—This technology introduces the concept of a switching hub to the Ethernet environment. The thrust of this approach is to deliver more of the possible 10 Mbps Ethernet bandwidth to each system in the network by reducing access contention between systems. Furthermore, because this approach is based on current 10 Mbps Ethernet technology, it allows the use of existing Ethernet cabling, adapters, and software.

☐ Full-duplex Ethernet—Under this approach, transmit and receive operations are assigned independent circuits and allowed to function concurrently between two systems. In theory this yields an effective transmission speed to 20 Mbps (10 Mbps in two directions). Full-duplex Ethernet is commonly used as an effective means of interconnecting bridges, repeaters, or hubs, but it can also be used to connect individual systems to one or more intelligent hubs.

☐ 100 Mbps Ethernet over UTP—The goal of this technology is to deliver Ethernet (or Ethernet-like) connections operating at speeds of 100 Mbps over Unshielded Twisted Pair (UTP) cable. This approach has engendered a fair amount of conflict and controversy, with several different factions offering different solutions with varying degrees of impact on the LAN environment. Depending on which solution you endorse, you may or may not be able to use existing UTP cable, and you may or may not be able to use existing Ethernet software drivers.

These approaches are not mutually exclusive. Many vendors are therefore devising products for multiple approaches, and some are even building devices that are a hybrid of these technologies.

Ethernet switching

The idea behind Ethernet switching is simple: in most networks, no single device ever comes close to enjoying the full 10 Mbps of theoretical Ethernet bandwidth. Therefore, if availability of existing bandwidth can be increased, performance will improve. Of course, this begs the question "why isn't the full 10 Mbps bandwidth (or close to it) available in the first place?"

Even in an ideal world, a system will never achieve 100% of Ethernet's 10 Mbps bandwidth because of network protocol constraints (e.g., delays caused by frame acknowledgments) and because of the electronics of the LAN (e.g., the time for the medium to "settle" after a transmission). For the sake of argument, let's say that in the worst case these two factors eat up 20% of the bandwidth. That means that in the simple case where only two systems are on an Ethernet LAN, they can only access 8 Mbps of bandwidth.

In most networks, however, multiple systems compete for access to the LAN. Given the contention architecture of the CSMA/CD access discipline, it follows that when more systems are put on a LAN, more delays will be experienced. Depending on the number of systems on an Ethernet LAN and the frequency with which they want to transmit, this can drive average performance down into the neighborhood of 1 Mbps (although an average of 3 to 6 Mbps is more typical). The bottom line is that performance degrades as you add more systems to an Ethernet network.

With this background in mind, the purpose of an Ethernet switching hub is to reduce contention for the 10 Mbps bandwidth by

☐ breaking an Ethernet LAN into multiple segments, with a small number of systems (1 to 12) on each physical segment.
☐ providing intelligent, high-speed switching between segments to ensure no delays are introduced when traffic flows from one segment to another.

The simplest configuration is the easiest to understand. Here, an Ethernet switching hub provides some number (say 8) of Ethernet segments, with a single system attached to each segment. The switching hub then acts as the control center and moves the traffic from a transmitting segment to the appropriate receiving segment at backplane speeds. In effect, this allows a transmitting and receiving system to enjoy performance similar to what they would experience if they were the only two systems on the LAN.

Because having a one-to-one relationship may seem like too heavy a penalty (as expressed in the number of hubs required to build a network), some switching hubs allow many (typically up to 12) systems to reside on each segment. The switching hub still acts as the control center, but each system must compete with other systems on the same physical segment. This cannot yield the same performance advantage offered by a hub with one system per segment, but the shared segment approach can certainly move performance from the low end of the scale toward the high end.

In large networks, one switching hub may not be enough, so multiple hubs will need to be interconnected. The critical factor in this situation is the speed of the connections between the hubs, so high speed (100 Mbps) fiber connections are normally used.

Full-duplex Ethernet

Implementation of full-duplex Ethernet requires that systems be wired in a point-to-point or star topology so that no more than two systems (or one system and a hub) share the same physical connection. This step eliminates competition for a common LAN medium, as is traditional in thick- or thin-wire Ethernet networks. Once systems are wired in this fashion, full-duplex Ethernet allows concurrent transmit and receive operations over the same physical connection, thus doubling the effective speed from 10 Mbps to 20 Mbps.

In reality, full-duplex operations are not effective or practical for end-user systems. These systems tend to transmit and then ex-

48

pect a response—they are not oriented toward transmitting data as part of one application function and receiving data as part of another. Network-oriented devices like hubs, repeaters, and bridges, however, are extremely well suited to full-duplex functions. As you can imagine, it is not uncommon for two hubs to want to send traffic to one another at the same time. When these types of network devices are interconnected using full-duplex Ethernet connections, an effective speed of 20 Mbps can, in fact, be realized between the two devices.

100 Mbps Ethernet over UTP

All of the Ethernet vendors agree that implementation of Ethernet (or Ethernet-like) connections at 100 Mbps is necessary to breathe continued life into Ethernet and position it as an alternative to running the Fiber Distributed Data Interface (FDDI) over shielded twisted pair cable. Unfortunately, that's about all they agree on.

The key differences between the various factions lie in two areas: UTP cable and access discipline. In order to understand these differences, however, we need a basic understanding of the cable types:

☐ UTP cable: UTP cable is available in five levels (termed "categories") of conditioning. All of the levels can carry voice transmission, so the primary difference is their rating for data transmission:
  ~ Level 1 and level 2 UTP cable handle data transmission at rates up to 4 Mbps.
  ~ Level 3 handles data transmission at rates up to 16 Mbps. This is the most common cable type used for combined voice and data traffic.
  ~ Level 4 UTP facilitates transmission rates up to 20 Mbps.
  ~ Level 5 accommodates data traffic at rates up to 100 Mbps.

UTP cables also can have different numbers of wires inside. Common configurations are 2 pair (4 wires), 3 pair (6 wires), and 4 pair (8 wires). For example, running 10 Mbps Ethernet over 10BASE-T today requires two pairs of wires.

The UTP wire types and the standard Ethernet/802.3 CSMA/CD access discipline are key factors that differentiate the various proposals (and implementations) of 100 Mbps Ethernet:

☐ 100BASE-TX: This proposal supports the use of the standard Ethernet CSMA/CD access discipline over two pairs of type 5 UTP cable.

49

☐ 100BASE-T4: This recommendation supports the use of standard Ethernet CSMA/CD access discipline over four pairs of type 3, 4, or 5 UTP cable.

☐ 100BASE-VG: This is the most radical faction of the three, because it proposes abandoning the CSMA/CD access discipline in favor of an new access discipline called the Demand Priority Protocol. This faction also recommends the use of four pairs of type 3 UTP wire for the physical connection. HP has also worked with IBM to develop a Token Ring version of 100BASE-VG called 100BASE-VG/AnyLAN.

Also note that a fiber-based proposal—100BASE-FX—defines the use of two-strand fiber optic cable for 100 Mbps Ethernet operation.

Finally, even though all of these approaches support the existing Ethernet frame format, they all dictate the use of new adapters and new interconnection devices (repeaters, hubs, bridges, etc.).

Which is for you?

The effect that Ethernet switching, full-duplex Ethernet, or 100 Mbps Ethernet operation over UTP cable has on you depends on whether you are trying to solve today's problem or tomorrow's problem:

☐ If you need increased speed in an existing Ethernet network, Ethernet switching and full-duplex Ethernet can deliver tangible performance improvements with modest cabling changes.

☐ If you are working on long-term networking strategy or planning new cabling installation, you should take a hard look at how you plan to use UTP cabling in support of 100 Mbps operation. It is, however, important to work with a vendor to determine the exact number and type of wiring pairs that will be required (but allocating *at least* four pairs of UTP level 5 wire per network connection is usually a safe bet).

Regardless of which option you choose, you should be able to achieve a perceptible—and perhaps dramatic—improvement over standard Ethernet performance.

Token Ring/802.5

Access discipline

As discussed in the introduction to Part II, two variations of Token Ring networks are commonly implemented: those following the IBM standard and those following the IEEE 802.5 standard. Unlike the differences between Ethernet and 802.3, the differences between IBM Token Ring and 802.5 are relatively minor. A full accounting of these differences is presented later in this chapter.

Both Token Ring networks can operate at 4 or 16 Mbps. The original implementation of a 4 Mbps ring used a frame size of approximately 2 kB. This speed and frame size, however, put Token Ring networks at a disadvantage when compared to the performance of Ethernet/802.3 networks, which operate at 10 Mbps and feature a frame size of 1500 bytes.

To compensate for the difference, the Token Ring frame size was subsequently increased to 4K for 4 Mbps rings, and operation at 16 Mbps was introduced, supporting frame sizes up to 16K. The increased frame size for 4 Mbps operations allowed Token Ring networks to come close to the performance of Ethernet/802.3 networks, and operation at 16 Mbps allowed it to exceed the performance of Ethernet/802.3 networks.

In terms of the access discipline used in a Token Ring network, the key element is a special message termed a "token." The token is passed from one system to another to give each system the opportunity to transmit. The token is a message three bytes long, and this message is a variation of the standard frame format used to carry messages. The token message is structured as follows:

- ☐ Frame format of a token message.
- ☐ Start Frame Delimiter—A one-byte pattern indicating the start of the frame.
- ☐ Access Control—A one-byte field used for control and maintenance functions in normal data messages. The fourth

bit of this field is the "token" bit—when set to one, the frame is a token.

☐ End Delimiter—A one-byte pattern signaling the end of the frame.

Information about the frame structures used by IBM Token Ring and 802.5 for data messages will be presented later in this chapter.

The flow of the token is a logical ring (regardless of how the network is physically cabled), so the token will always end up back at the originating system. A system with information to transmit on the network must wait to receive a token. Once it has possession of the token, it can transmit a frame of data to another system. When the receiver obtains the data, it sets a flag in the frame acknowledging receipt and releases the frame back into the ring. The originator sees that the frame has made it (or not), and it generates a new token to allow another system to have access to the ring.

Whereas the CSMA/CD access discipline used in Ethernet/802.3 LANs resembles casual group conversation, the token passing discipline used in Token Ring networks parallels a highly structured, round-robin panel discussion. In this model, the first panel member is given the opportunity to make a statement and then control of the floor passes on to the next member. That member speaks and then relinquishes control to the next panel member. Finally, after the last panel member has spoken, control passes back to the first member.

The strength of the Token Ring methodology is that it guarantees that every system on the LAN will have the opportunity to transmit. This methodology also means that Token Ring behavior is extremely predictable—tokens and data frames travel from one system to another in a logical and orderly fashion. In larger networks, however, the strength of the Token Ring LAN can become a weakness. Specifically, each system must wait to receive a token before it can transmit; when more systems are added to a LAN, this increases the opportunity for transmission from multiple systems, and therefore can decrease the opportunity any one system has to receive a token. One compromise on this issue is to divide a large ring into several small rings. However, dividing systems into smaller rings is really only practical if traffic on the smaller rings can be self-contained. In other words, breaking a large ring into smaller rings which constantly access one another is not a dramatic improvement.

Ring monitoring

The overall flow of frames through the ring is regulated by an active ring monitor. Any system can potentially be an active ring monitor, although only one active monitor may be in effect at any one time. The responsibilities of an active monitor include

☐ looking for data frames that travel around the ring more than once—The active monitor will remove any offending frames from the ring and discard them.
☐ detecting the loss of a token—In this case the active monitor initiates a purge of the ring and then originates a new token.
☐ clocking and timing—The active monitor is responsible for providing the master synchronization of the ring. This is the clock source used by all other systems to ensure that they are using the same timing to send and receive data.

The loss of a token or the perpetual circling of a data frame are "soft" errors. They can be induced by accidental disruptions in the LAN, such as kicking a cable off (and then reconnecting it), or by powering down systems when they are in possession of a token or involved in a message exchange. In all of these situations, the active ring monitor should resolve the problem automatically.

An active monitor is elected. When a system detects the absence of a ring monitor or a failure of a ring monitor, it initiates an election procedure known as "token claiming." When one system detects the absence of the active monitor (based on the absence of a ring monitor present message), it initiates a "claim token" message which, in effect, is a request to become the active monitor. When another system receives the claim token message, it takes one of the following actions:

☐ If the system does not want to participate in the election process, it simply passes on the message.
☐ If the system wants to participate, but the address of the system that generated the claim is higher than its own, it passes the message on.
☐ If the system wants to participate and has a higher address than the system that generated the claim, it initiates a new claim token message.

In the end, the system with the highest address participating in the claim procedure will end up being elected the new active monitor.

53

Ring access

Individual systems are responsible for testing and participating in ring integrity checking. The bulk of this testing is performed when a system accesses a ring for the first time, and involves five specific phases:

☐ Phase 0 (Lobe Test)—In this phase the system sends a series of test messages to itself in order to test the connection between itself and the Multistation Access Unit (MAU). If this phase fails, the system will not proceed any further.

☐ Phase 1 (Monitor Check)—During this phase the attaching system "listens" to the ring to see whether an active monitor is present. As noted, an active monitor is a system that assumes responsibility for resolving correctable problems on the ring. An active monitor indicates its presence by periodically transmitting "active monitor present" messages on the ring. If the attaching system does not receive a broadcast from an active monitor, it initiates the procedure used to elect an active monitor (or it assumes the role of the active monitor if it is the only system on the ring). Once the operation of an active monitor is resolved, the system proceeds to the next phase.

☐ Phase 2 (Duplicate Address Check)—In this phase the system initiates a test message that checks for duplication of its own address in another system. If a duplicate address is detected, the system stops communicating with the ring.

☐ Phase 3 (Neighbor Notification)—During this phase the attaching system sends test messages to learn the address of its Nearest Active Upstream Neighbor (NAUN) and to notify its downstream neighbor of its address. The concept of NAUN is important in deciphering Token Ring problems. Specifically, when a serious problem occurs in the ring, the system that detects the problem starts complaining and identifies its upstream neighbor as the probable cause. This mechanism is called *beaconing* and will be discussed later.

☐ Phase 4 (Request Initialization)—The final phase is used to receive any special parameters implemented through a ring parameter server. This might, for example, include software level IDs or ring number assignment. The use of initialization parameters is optional (although all systems should make the request in this phase).

These access-time tests and the activities of the ring monitor are the two key areas that facilitate the automatic monitoring and resolution of many Token Ring network problems.

Problem notification

When a problem arises that cannot be corrected by the active ring monitor or by the individual system that detected the error, the problem is escalated to a higher level through beaconing. The intent of beaconing is to isolate the malfunctioning system and to inform the human operator (or any network management products that might be running) that a serious and uncorrectable problem is occurring.

A beacon is a special type of network message originated by the system that first detects the problem. The beacon message identifies the system reporting the problem and also identifies the address of its Nearest Active Upstream Neighbor (NAUN). This information is included in the beacon message to help isolate the problem—in theory, if a system is causing problems on a LAN, these problems will first be detected by its downstream neighbor.

If the beaconing system's NAUN is capable of recognizing the beacon message, it will watch for eight beacon messages to pass by and then it will take itself off the ring and perform self-tests. These self-tests correspond to the Phase 0 (Lobe Access Test) and Phase 2 (Duplicate Address Test) checks performed during the normal attachment process. If the NAUN successfully completes these tests, it assumes the problem has been corrected and rejoins the LAN. If, however, these tests do not complete successfully, the NAUN remains off the ring.

The beaconing system itself will wait for a period of time (based on ring size and speed) for the problem to be resolved. If the problem remains after the timer expires, the beaconing system will assume the NAUN has taken appropriate action. At that point the beaconing system will take itself off the LAN and perform the Lobe Access Test and Duplicate Access Test. If the tests are successful, the system rejoins the LAN. If unsuccessful, the system remains detached from the ring.

If a system is experiencing an internal hardware problem that can be detected but not corrected, the beaconing process will identify it as the NAUN, it will perform self-tests, and it will detach from the ring and remain detached until human intervention corrects the problem.

If, however, a system is exhibiting radically deviant behavior on the LAN, but the hardware itself cannot detect that it has a problem, the beaconing process will be repeated over and over until human intervention occurs—the defective system will be identified by the beacon and perform a successful self-test, as will the beacon, but the problem will continue. Needless to say, this has a negative impact on normal LAN activities.

Unfortunately, the latter case is more prevalent than the former—most hard errors are detected and resolved by humans, not by software.

Optional behavior

In the token passing methodology previously described, each system has equal access to the token—no one system has priority over another. An option in most Token Ring implementations is the use of multiple access priorities for systems on the LAN. These priorities may be applied to LAN traffic to allow higher-priority systems to "override" the transmissions of lower-priority systems. In effect, one system can interrupt the current traffic on the LAN to handle higher-priority traffic.

When a system has a message to transmit but senses a data frame instead of a token, it compares its own priority to the priority of the current data frame. If the system wishing to transmit has higher priority, it sets a flag in the data frame indicating that it wants to interrupt the current flow of data. When the data frame returns to its point of origin, the system notes that a higher-priority request has been received. Instead of generating a standard token, the lower-priority system generates a higher-priority token that corresponds to the priority level of the system requesting the interrupt.

This high-priority token then makes its way to the interrupting system following the normal flow of data. Systems that see the token will not seize it unless they have an equal or higher priority than the system that requested the interrupt. Once the high-priority token is received by the interrupting system, it transmits data at that priority level. When the system is done with its priority transmissions, it generates a new token with a priority level that corresponds to the original interrupted token. At that point the original flow of information is restored.

Another option in Token Ring LANs is early token release. When this method is employed, a system originating a data frame releases a token immediately after the transmitting data frame, instead of after receiving the data frame on its return trip. This approach allows multiple data frames to be present on the ring at the same time, thereby facilitating greater throughput.

Addressing scheme

Once a system gains possession of the token in an IBM/802.5 Token Ring LAN, it can transmit a frame of information over the LAN. The general elements that go into a LAN frame under the IBM/802.5 specifications are the following:

☐ Control flags to indicate whether the frame is a token, a network protocol (data) message, or a Token Ring control message. If the control flag indicates the message is a token, the remaining information is omitted.

☐ A source address that identifies the system originating the frame.

☐ A destination address that identifies the system (or group of systems) the frame is intended for.

☐ An indication of what network protocol (e.g., TCP/IP, IPX, SNA, DECnet) is carried in the frame.

☐ A data area that contains the network protocol information along with any "real" data. This network protocol information can also contain high-level addressing information that defines a logical address and logical network for the transmitting and receiving systems.

☐ A checksum that enables the receiver(s) to verify that the frame has arrived intact.

☐ A status flag to indicate whether the frame was received by its intended receiver(s) or not.

The source and destination addresses correspond to low-level (MAC-level) hardware addresses of the network interface card, or addresses overlaid on top of the hardware addresses. As in the case of Ethernet/802.3, these addresses are the "bottom line" for getting a frame to a system.

Token Ring uses a 48-bit address scheme like the Ethernet/802.3 scheme. The address is represented as six one-byte hexadecimal values. For example, 00-00-F6-08-08-B1, 10-00-5A-D3-B1-7C, and 42-60-8C-0B-69-16 are valid Token Ring addresses. Each system must have a unique address to participate in a Token Ring network. That address is normally assigned by the hardware manufacturer and programmed directly into the network adapter hardware, but

57

in most Token Ring networks, the address is overlaid with a "soft" address, termed the *locally administered address*.

The hardware-assigned addresses follow the IEEE address format, wherein each manufacturer is assigned a three-byte address (or range of addresses) that initiates the six bytes of address information. Please refer to Table 2-1 in Chapter 2 for a complete list of manufacturer assignments.

Oddly enough, in a Token Ring network address, the bits in each byte of the network address are arranged in reverse order from an Ethernet/802.3 address. For example, the hexadecimal value 02 (binary 0000-0010) in an Ethernet/802.3 address turns into the value 40 (binary 0100-0000) in a Token Ring address. This transformation affects all aspects of the address. Thus, the assigned IBM address prefix 08-00-5A (binary 0000-1000 ¦ 0000-0000 ¦ 0101-1010) becomes 10-00-5A (binary 0001-0000 ¦ 0000- 0000 ¦ 0101-1010). Similarly, the Madge Networks prefix of 00-00-6F (binary 0000-0000 ¦ 0000-0000 ¦ 0110-1111) becomes 00-00-F6 (binary 0000-0000 ¦ 0000-0000 ¦ 1111-0110).

Although IBM has presented countless arguments as to why this format is faster and easier to implement, most of the industry views it as bizarre, or blindly accepts it, or both.

When the preset hardware address is not used, a locally administered address must be set up. A locally administered address is enabled by setting bit 0 to "0" and bit 1 to "1" in the first byte. For example, if the first byte of the address is 00, the result would be 02 (binary 0000-0010), but since this is a Token Ring address, the bits are flipped and the resulting address byte is 40 (binary 0100-000). The remaining portion of the address may be set to any desired value. The end result is that any address in the range 40-00-00-00-00-00 to 7F-FF-FF-FF-FF-FF (after bit-flipping) is a "legal" locally administered address.

In addition to specific system addresses, Token Ring also supports group and broadcast designations in destination addresses that follow the Ethernet/802.3 convention. Thus, Token Ring supports an all-systems broadcast address of binary ones (FF-FF-FF-FF-FF-FF) as well as "multicast" addresses, which are designated by setting bit 0 in the first address byte. (Don't forget about bit-flipping, however, so an Ethernet/802.3 multicast address that begins 01 will become 80 in a Token Ring address.)

Whereas bit 0 is the multicast bit in the destination address value, it has a totally different meaning in a source address value. Specifically, bit 0 in a source address is the source routing indicator. If bit 0 is set to "1," this indicates that the transmitting system uses source routing to locate systems not on the local ring. Systems that do not use source routing must set this bit to "0."

Token Ring LANs also make use of another type of address, termed a *functional* address. These addresses are used to perform Token Ring operational functions, such as controlling the ring monitor function. Functional addresses set bits 0 and 1 to "1" (resulting in a Token Ring value of C0) and also set bit 0 in the second address value to "0." For example, address C0-00-00-00-00-01 is the functional address of the active ring monitor, and address C0-00-00-00-00-02 is the functional address of the ring parameter server.

When it comes to setting addresses in systems attached to a Token Ring LAN, the best advice is to forget about the bit-flipping that occurs and simply start each address with hexadecimal 40 (or "4x") to make it a locally administered address. You can then set the remaining five bytes in the address as you like, and since your interface to this address will normally be at a high level, you will be insulated from the bit-flipping that is occurring at a lower level.

You may not, however, be able to avoid the issue of bit-flipped addresses all of the time—for example, you will be confronted with this issue if you bridge Ethernet/802.3 and Token Ring LANs—but for normal operations, the farther away you stay from this issue, the happier you will be.

Source routing

Source routing is a technique that enables a system located on one physical ring to communicate with a system located on a different physical ring. Source routing is defined in both the IBM and IEEE token-ring specifications (although the details vary slightly between them). In reality, many networking vendors do not use source routing, relying instead on higher-level (network protocol-level) functions to handle LAN-to-LAN routing. IBM, however, makes heavy use of source routing in their LAN products.

Source routing depends on the presence of bridges to interconnect physically separate rings. In a real sense, bridges are the key component to the source routing technique, because the bridges provide the route discovery process that enables one system to determine how to reach another system. As this discussion implies, bridges used in a source routing environment must explicitly

59

support this technique. Please refer to Chapter 4 for further discussion on bridges and how source routing affects bridging functions in multivendor environments.

The principle of source routing is really a relatively simple extension to the way that systems in a Token Ring LAN initiate communication with one another. Specifically, when one system wants to communicate with another, it first sends a "test" message to the other to ensure that it is on the ring and is in an operational state. If that system is present, it responds to the test message and the two systems can initiate a dialog.

Source routing kicks in when the system does not receive a response to a test message. In this case, the initiating system turns on the source route bit (bit 0 in the source address field) and sends another test message. Any source routing bridges attached to the same ring as the initiating system "see" the request and send the test message to the next ring they are attached to, adding some simple routing information to the frame in the process. This message is then propagated to all rings in the network.

When the test message actually reaches the intended system, the frame contains a description of the route it took to get there (e.g., which bridges it had to cross). The receiving system then returns an acknowledgment to the initiating system and the routing information remains in the frame so the initiating system knows how to reach the system. At that point the two systems can communicate with one another as they see fit.

The routing information generated in a source-routing environment is inserted into the standard Token Ring frame immediately after the source address. This information is composed of the following fields:

☐ A two-byte routing control field. This field indicates, among other things, whether the frame is a route discovery message or the route has already been determined. This field also specifies the number of route designator fields that follow.
☐ Two or more route designator fields follow the routing control component. (As discussed later in this chapter, the IBM and IEEE specifications for the number of allowed designators differ.) A route designator field is two bytes long. The first 12 bits identify a logical ring number, and the remaining four bits pinpoint a bridge number where the message traveled through. The last route designator field identifies the ring where the receiving system is located, and has a bridge number of "0."

As shown in Fig. 3-1, two route designator fields will be present in a frame that has traveled over one bridge—the first defines the ring number where the frame originated and the bridge number it traveled through. The second route designator field defines the ring number of the target system.

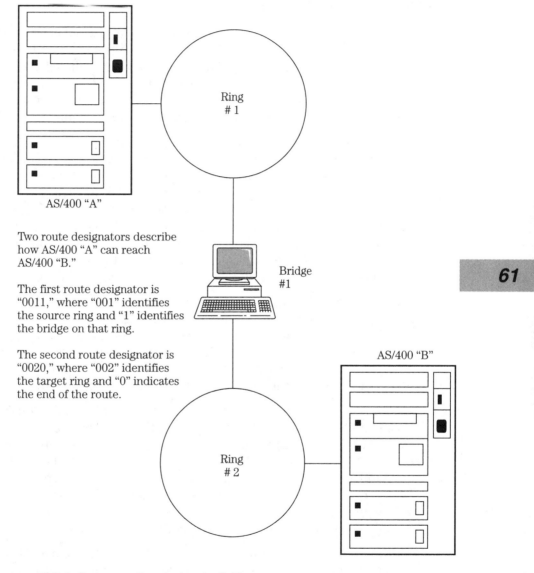

AS/400 "A"

Two route designators describe how AS/400 "A" can reach AS/400 "B."

The first route designator is "0011," where "001" identifies the source ring and "1" identifies the bridge on that ring.

The second route designator is "0020," where "002" identifies the target ring and "0" indicates the end of the route.

Bridge #1

AS/400 "B"

■ **3-1** *Source routing designator fields*

Systems using source routing can coexist with systems not using source routing, but they cannot use the same type of bridge or router for wide area connections. Please consult Chapter 4 for more information on this topic.

Topology & construction

The original Token Ring networks were, in fact, wired in a physical ring. Cables were daisy chained from one system to another until they formed a closed loop. This was not a great architecture, however, because any break in the ring disabled the entire LAN—if you disconnected a PC, you disconnected the LAN. Given this limitation, very few physical ring networks have survived into the 1990s.

As shown in Fig. 3-2, today's Token Ring LANs are cabled in a hub configuration. In this configuration, systems are attached via a cable (termed a "lobe") to a central device called a Multistation Access Unit (MAU), which supports a number of lobe connections. For example, the IBM 8228 MAU can support up to eight lobe connections. Every Token Ring network must have at least one MAU. If the number of required lobes cannot be accommodated by one MAU, multiple MAUs (up to 33) can be connected together using dedicated *Ring In* (RI) and *Ring Out* (RO) connections on each MAU.

When multiple MAUs are used, the distance between MAUs is dependent on the medium. If conventional Type 1 cable is used, an MAU can be placed within 100 meters of the adjacent MAU. Longer distances may be achieved through the use of copper repeaters, which extends that distance to 740 meters, or fiber repeaters, which can extend the distance up to 4 kilometers.

A lobe connection from an MAU often normally attaches directly to a system. If multiple systems are at that location, a *Lobe Access Unit* (LAU)—sometimes also referred to as a hub—might be used to split the single lobe into two or more lobes. Using this scheme, an LAU can be placed on a single cable drop to accommodate the connection of two or more systems. Many LAUs also support a dedicated attachment to accommodate another LAU, so LAUs can be chained together to provide additional attachments.

In theory, the difference between an LAU (or hub) and an MAU is that an LAU cannot be used alone to create a stand-alone ring. Unfortunately, different manufacturers use the terms *LAU* and *hub* loosely, so some LAUs are in reality MAUs. If you are purchasing an LAU or a hub, you should find out whether it can be used as a stand-alone device.

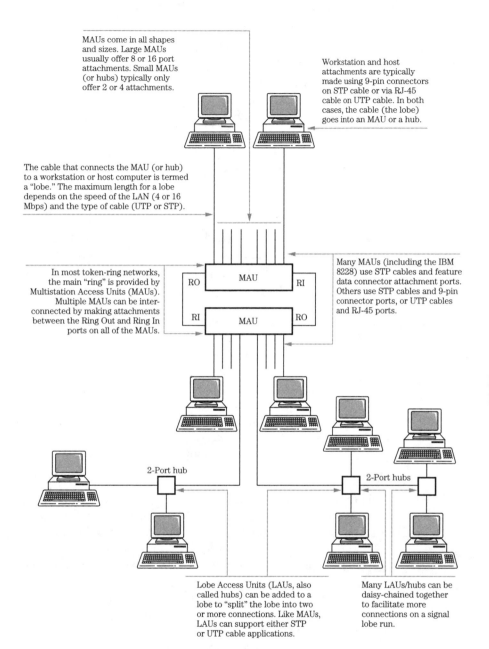

MAUs come in all shapes and sizes. Large MAUs usually offer 8 or 16 port attachments. Small MAUs (or hubs) typically only offer 2 or 4 attachments.

Workstation and host attachments are typically made using 9-pin connectors on STP cable or via RJ-45 cable on UTP cable. In both cases, the cable (the lobe) goes into an MAU or a hub.

The cable that connects the MAU (or hub) to a workstation or host computer is termed a "lobe." The maximum length for a lobe depends on the speed of the LAN (4 or 16 Mbps) and the type of cable (UTP or STP).

In most token-ring networks, the main "ring" is provided by Multistation Access Units (MAUs). Multiple MAUs can be inter-connected by making attachments between the Ring Out and Ring In ports on all of the MAUs.

MAU
RO   RI

RI   RO
MAU

Many MAUs (including the IBM 8228) use STP cables and feature data connector attachment ports. Others use STP cables and 9-pin connector ports, or UTP cables and RJ-45 ports.

2-Port hub

2-Port hubs

Lobe Access Units (LAUs, also called hubs) can be added to a lobe to "split" the lobe into two or more connections. Like MAUs, LAUs can support either STP or UTP cable applications.

Many LAUs/hubs can be daisy-chained together to facilitate more connections on a signal lobe run.

■ **3-2** *Token Ring construction*

63

The Token Ring approach to cabling accommodates all sizes of networks:

☐ As shown in Fig. 3-3, a two-system ring can be constructed with a single 2-port MAU.
☐ As also shown in Fig. 3-3, complex networks can be constructed by chaining 8-port MAUs to one another and optionally attaching additional 2-port hubs to lobes leading from the 8-port hub. Notice that the 8-port MAUs are connected together using *Ring In* (RI) and *Ring Out* (RO) cables to extend the physical ring over multiple MAUs.
☐ MAUs and LAUs maintain the integrity of the ring. When a lobe connection is made to an MAU or an LAU, it "opens" the ring to facilitate the new attachment. Similarly, when a lobe is disconnected from an MAU or an LAU, the connection point is automatically "closed" to keep the ring intact. This flexibility allows the construction of the network to be changed without bringing it down.
☐ Token Ring networks are often implemented as a set of interconnected rings. When this approach is used, a bridge is used to tie the two networks together.

Cabling

The physical construction of a Token Ring LAN is dependent on the medium as well as the placement of MAUs and hubs. Token Ring networks are generally constructed out of two types of cable media:

☐ Shielded Twisted Pair (STP) cable (also known as *IBM Type 1*)—STP cable is composed of shielded twisted pair strands and is suitable for lobe connections up to 100 meters long. STP cable is terminated in either a 9-pin D-shell connector or a "patch" connector. As a general rule, patch connectors are for MAU attachments, male 9-pin connectors are for system and LAU attachments, and female 9-pin connectors are used to originate a daisy-chain attachment from one LAU to another. STP cable is available for indoor and outdoor environments. IBM Type 1 cable calls for two shielded pairs.
☐ Unshielded Twisted Pair (UTP) cable (also known as *IBM Type 3*)—UTP cable is composed of unshielded twisted pair strands similar to telephone cable. The connectors used with Type 3 cable are RJ-45 modular plugs. UTP cable may be used for lobe connections in the neighborhood of 45 meters long (the actual length depends on the speed of the LAN and the

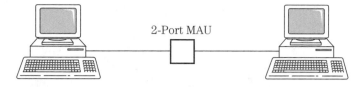

2-Port MAU

As shown above, small token-ring networks can be constructed using a single MAU or hub. Multiple MAUs and hubs can also be interconnected to create large networks, as shown below.

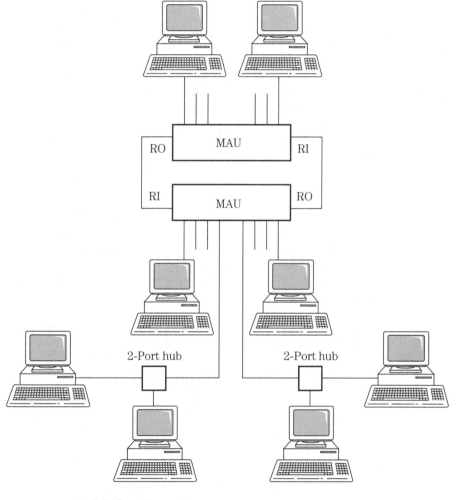

■ **3-3** *Small/large rings*

characteristics of the environment). IBM Type 3 cable calls for four unshielded pairs.

IBM also supports the use of Type 2, Type 5, Type 6, Type 8, and Type 9 cable in Token Ring networks. These cable types are, however, specialized for certain environments or applications:

☐ Type 2 is a combination of Type 1 (shielded twisted pair) and Type 3 (unshielded twisted pair) strands. This means that Type 3 cable includes two shielded pairs and four unshielded pairs.
☐ Type 5 is a 62.5/125 multimode fiber optic cable. In Token Ring environments this is often used to interconnect repeaters.
☐ Type 6 is a low-cost, short-distance (maximum of 45 meters) shielded twisted pair cable often used for MAU-to-MAU connections. Type 6 cable normally contains two twisted pairs.
☐ Type 8 is for sub-carpet installation and can run up to 50 meters. Type 8 cable is not composed of twisted pairs—it is composed of two parallel pairs instead.
☐ Type 9 cable is a shielded twisted pair cable that can run up to 65 meters. Type 9 cable contains two twisted pairs and is a lower-cost (and shorter-distance) alternative to Type 1 cable.

Token Ring uses the differential Manchester encoding scheme to transmit and receive data on the LAN. In both STP and UTP cables, signals are carried over two balanced circuits, one for transmit and one for receive. Each circuit is then made up of two wires, one used to carry the positive signals and the other used for the negative signals. This means that Token Ring cable requires two twisted pairs of cable.

When STP cable is used, the pin assignments for the two twisted pairs are as follows:

■ STP pin assignments.

Signal	9-pin	Patch
Transmit −	5	Black
Receive +	1	Red
Receive −	6	Green
Transmit +	9	Orange

Figure 3-4 shows the locations of these circuits on these two connector types.

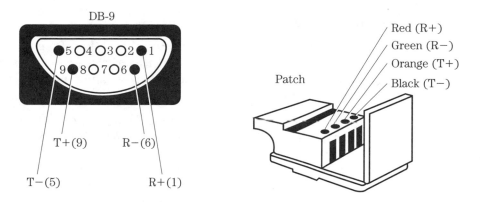

T+(9)   R−(6)

T−(5)   R+(1)

Red (R+)
Green (R−)
Orange (T+)
Black (T−)

Patch

■ **3-4** *STP/patch panel connectors*

When UTP cable is used, the wires are normally assigned to a standard RJ-45 connector as follows:

■ **Typical 10BASE-T/UTP cable.**

Wire	Assignment	Type 3 Wire Color
1		
2		
3	Transmit −	Blue with White stripe
4	Receive +	White with Orange stripe
5	Receive −	Orange with White stripe
6	Transmit +	White with Blue stripe
7		
8		

As shown in Fig. 3-5, wires 3 and 6 should be twisted together, and wires 4 and 5 should be twisted together. Note that this wiring is different from Ethernet over 10BASE-T, which uses the 1/2 and 3/6 pairs.

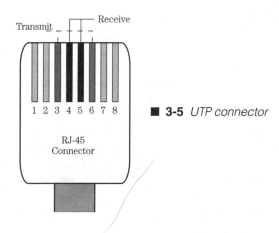

■ **3-5** *UTP connector*

67

IBM Token Ring vs. 802.5

Although the differences between the IBM Token Ring specifications and the 802.5 standard are not nearly so dramatic as the differences between Ethernet and 802.3, two differences are particularly important:

☐ IBM specifications state that a ring can support up to 260 attachments, while the IEEE 802.5 specifications limit support to 250.

☐ IBM specifications allow up to eight route designator fields when source routing is employed. The IEEE 802.5 specifications allow a maximum of 14.

With the exception of the maximum number of route designator fields allowed, the IBM and IEEE frame formats have the same field structure, as shown in Fig. 3-6.

The fields within the IBM/IEEE 802.5 frame formats are as follows:

☐ Start Frame Delimiter—A one-byte pattern indicating the start of the frame.

☐ Access Control—A one-byte field used for control and maintenance functions. The fourth bit of this field is the "token" bit. If set to "1," the frame is a token and is composed of just the Start Frame Delimiter, Access Control, and End Delimiter fields (total of three bytes long).

☐ Frame Control—A one-byte field used to identify specific information, control, and maintenance functions.

☐ Destination Address—The address of a specific station or group of stations to receive the frame. May be two or six bytes long.

☐ Source Address—The address of the sending station. May be two or six bytes long.

☐ Routing Control—This field is only present when source routing is enabled (the first bit of the Source Address field is set to "1"). When present, this field specifies whether route discovery is required, and how many Route Designator fields follow.

☐ Route Designators—Route Designators are also only present when source routing is enabled. A minimum of two fields will be present. Each two-byte route designator defines a ring number and a bridge number that the frame passes through. The last designator defines the ring number of the receiving system and has a bridge number of 0. As noted, IBM allows up to eight Route Designator fields while the IEEE specification sets the limit to 14.

Bit 0 of the source address (which is shown as the left-most bit according to token-ring specifications) is the source routing bit. When set to "1," source routing information is inserted into the frame.

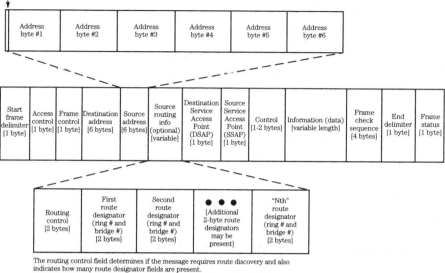

The routing control field determines if the message requires route discovery and also indicates how many route designator fields are present.
Each route designator field records a ring number and a bridge number that the message must travel through to reach its destination.

■ **3-6** *Token Ring frame layout*

☐ DSAP—The one-byte 802.2 Destination Service Access Point. See the discussion under the *IEEE 802.2* heading in the introduction to Part II for more information.

☐ SSAP—The one-byte 802.2 Source Service Access Point. See the discussion under the *IEEE 802.2* heading in the introduction to Part II for more information.

☐ Control—An 802.2 field used for various commands, including *exchange identifier*, *test*, *connect*, *disconnect*, and *frame reject*. May be one or two bytes long, with the length defined by the first two bits.

☐ Data—The actual information being transmitted.

☐ Frame Check Sequence—A four-byte field containing a checksum of the frame information beginning with the Frame Control field and ending with the data.

☐ End Delimiter—A one-byte pattern signaling the end of the frame.

☐ Frame Status—A one-byte field used for status information. Specifically, this field is set by a receiving station to indicate that it has recognized and copied a frame addressed to it.

Overall specifications

General

Speed	4 or 16 Mbps
Maximum frame size	4511* bytes (4 Mbps), 17,839 (16 Mbps)
Maximum data unit	4414* bytes (4 Mbps), 17,742 (16 Mbps)

*The original implementation of 4 Mbps Token Ring was limited to a maximum frame size of 2091 bytes with a 1994-byte data unit size.

Type 1 (STP) medium

Connector	DB-9 or Patch
Maximum lobe length	100 meters
Maximum attachments	260 (IBM) 250 (802.5)

Type 3 (UTP) medium

Connector	RJ-45
Maximum lobe length	45 meters
Maximum attachments	72

High-speed Token Ring alternatives

Like Ethernet/802.3 networks, IBM/802.5 Token Ring networks have several options for upgrading beyond the current commercial speed of 16 Mbps. The three most realistic options include

☐ Token Ring operation at 32 Mbps (or higher)—IBM has been testing the Token Ring access discipline at higher speeds over both existing and higher-speed cable types. Although commercial implementations of higher-speed Token Ring solutions have yet to corner the Token Ring market, operation at 32 Mbps seems rational and reasonable, and operation at up to 64 Mbps may also be obtainable.

☐ Token Ring switching—Although Token Ring networks cannot be switched in the same sense that Ethernet networks can be, several vendors have introduced "switching" bridges to speed up the way that messages travel between rings. These devices generally use one of two techniques. The first is to start switching a message as soon as the address is recognized. This technique is called *cut through*, and it provides better performance than buffering the entire message and switching it as a whole. The second is *full duplex* Token Ring, which implements concurrent transmit and receive operations. As in the case of Ethernet, full duplex Token Ring is best suited to operate between bridges or other interconnection devices.

☐ "Low speed" Asynchronous Transfer Mode (ATM)—IBM has pioneered a low speed ATM implementation that operates at 25 Mbps over Type 3 UTP cable and uses the same physical medium interfaces as Token Ring. Since each connected user gets full access to the 25 Mbps of bandwidth, low speed ATM provides higher performance than Token Ring without requiring reinvestment in cabling.

☐ Fiber Distributed Data Interface (FDDI)—FDDI is not "officially" in the Token Ring family of LANs, but it has a great deal in common with it. Under FDDI, systems are arranged in a fiber-based ring (dual rings are deployed for redundancy), and a token-passing access discipline is followed. FDDI is certainly closer to Token Ring than it is to Ethernet/802.3. More information on FDDI is presented in the second half of this book.

☐ 100BASE-VG/AnyLAN—The same UTP-based 100 Mbps network solution available for Ethernet/802.3 networks is also available for Token Ring networks. Under this implementation, systems use a token-passing protocol instead of a demand-access protocol. Please see the discussion under the *100 Mbps Ethernet Over UTP* heading in Chapter 2 for more information on this subject.

As in the case of most of the high-speed Ethernet/802.3 options, upgrading to a higher-speed Token Ring alternative requires new adapters and interconnection equipment.

71

Interconnection tools

TO PARAPHRASE JOHN DONNE, "NO LAN IS AN ISLAND" IN THE world of modern networking. No matter how narrow the scope or vision of an initial LAN implementation, sooner or later it will reach a point where it needs to expand into distant geographical areas (e.g., across the street or across the globe) or connect up with other networks. And that's where the complications start.

In some cases, the need to connect two LANs together will be relatively straightforward—for example, two groups of users on two separate LANs may need to access a common database server. In others, the interconnection will be far more complicated—several PC users on a LAN may need to access a mainframe SNA network, or a PC-based program may need to retrieve information from a remote Unix system. The simple truth is that there are hundreds of scenarios that can break down the walls of "self-contained" LANs.

When those LAN walls come tumbling down, new tools are required to make network-to-network connections. A variety of tools are available, and each tool has its own unique options and considerations. In the most general terms, however, there are four basic interconnection tools:

☐ Bridges—Bridges are simple networking devices that interconnect two or more LANs. Bridges operate at the lowest network level and are not aware of what networking protocols are in use. For example, a typical bridge does not "see" any substantial difference between an Ethernet frame containing

Novell IPX information and an Ethernet frame carrying DECnet information—they are both simply Ethernet frames to the bridge.

☐ Routers—Routers are sophisticated networking devices that interconnect and regulate traffic flow between two or more LANs. Unlike bridges, routers operate at the network protocol level and can distinguish one protocol from another. In fact, understanding network protocols is the key technology to routers, because routers move information based on the protocol, or based on information carried by the protocol. For example, a LAN-based router may route Novell traffic over a wide area link and ignore all of the other protocols in the LAN.

☐ Gateways—In the broadest possible sense, gateways serve to interconnect dissimilar networks. Exactly how they accommodate interconnection and what specific dissimilarities they address can be very different from one gateway to the next, and this makes gateways hard to define. To a certain extent this makes gateways the "catch-all" category for network products—if an interconnection device can't clearly be labeled a bridge or a router, it will probably be called a gateway. For example, one gateway may permit PC users to access a remote IBM mainframe network, while another accommodates file and print transfers between an IBM AS/400 network and a Digital VAX LAN.

☐ Hubs—Like gateways, hubs are hard to define because they are available in a number of configurations, with each configuration offering a different level of service. At the low end are "dumb" hubs that act as LAN segment repeaters or wiring hubs. At the other end of the spectrum are "smart" hubs that provide sophisticated wiring management, support network management protocols (e.g., SNMP), and offer bridging, routing, and gateway functions.

As shown in Fig. III-1, we can look at these four types of tools from the perspective of the Open Systems Interconnect (OSI) model and see that bridges, routers, and "dumb" hubs have well-defined areas of responsibility, while gateways and "smart" hubs run the entire gamut of the model. If, however, we look at these four tools from a less analytical (and certainly more practical) point of view, each of these tools is a key piece in the larger puzzle of internetworking:

☐ Bridges provide the most economical means of interconnecting two or more LANs. Because bridges operate independently of network protocols, they are universal in

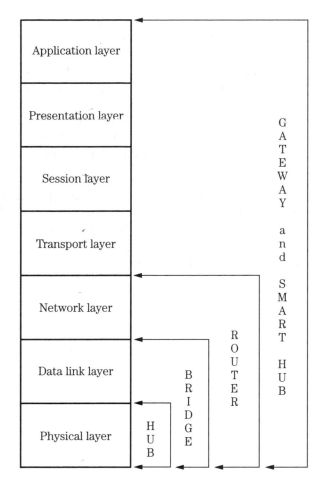

■ **III-1** *OSI model and bridges/routers/gateways*

nature. On the negative side, bridges tend to let extraneous network traffic cross over from one network to another.

☐ Routers provide the sophistication you need for interconnections between multiple LANs running multiple protocols. Routers are intelligent, and are active participants in the network protocols they handle (e.g., TCP/IP or IPX). Unlike bridges, routers enable you to regulate traffic flow across interconnections, and also unlike bridges, routers cannot be used with the entire range of network protocols.

☐ Gateways solve the "hard-to-reach" problems that crop up when one type of network traffic needs to cross over or interact with a different type of network. You can safely assume that some form of gateway is available for virtually every unique network-to-network connectivity requirement that cannot be solved by a router or a bridge.

☐ Hubs minimize the countless headaches that arise when you implement or expand LAN wiring. The value that a hub offers depends on where you are in the implementation cycle. If you are planning a new network or expanding an existing one, hubs offer features and benefits that easily outweigh the cost-savings of a "do-it-yourself" wiring system. If, on the other hand, your network is installed and stable, hubs aren't nearly so attractive.

As you can imagine, each of these four types of tools has its own unique attributes and technical considerations. To help clarify these issues we will explore the properties of bridges and routers in Chapter 4, and then in Chapter 5 we will examine the characteristics of gateways and hubs.

Bridges & routers

THE WORD "LOCAL" IN "LOCAL AREA NETWORK" IMPLIES that LANs have specific limitations on the geography they can cover and the number of attachments they support. As discussed in Chapter 2, the Ethernet specifications define limits for segment sizes, for the overall length of all combined segments, and for the number of overall attachments. Similarly, the Token Ring specifications presented in Chapter 3 define limits for individual lobe lengths, for the overall length of the ring, and for the number of overall attachments. When network requirements exceed these specifications, multiple LANs must be employed.

LAN specifications are not the only driving force for implementing multiple LANs—multiple LANs may also be established for geographic, functional, or political purposes. For example, a large corporation may have one LAN in Boston and a separate one in San Francisco. A second corporation may implement separate accounting and engineering LANs for optimum security and performance. Or yet another company may implement independent Novell NetWare, IBM SNA, and TCP/IP LANs. As you can imagine, the scenarios are almost endless.

Regardless of how multiple LANs come into creation, rarely can they stay separate and independent from one another for long. Somehow the march of corporate business invariably creates a requirement (or crisis) to tie some or all of the separate LANs together—implementing corporate-wide electronic mail, high-quality (or high speed) network-based printers, centralized databases, and many other common practices often dictate that workstations and systems access resources on LANs they are not directly attached to.

Without question, the easiest way to join multiple LANs is to physically interconnect them and make one big LAN (providing, of course, the combined LANs don't exceed the specifications for overall length or maximum number of attachments). For example, if an accounting Ethernet LAN is a couple of feet away from an engineering LAN, a simple connector could be installed to join them together. The downside of this solution, however, is the dramatic increase in traffic (and corresponding decrease in response time) for each department, because suddenly accounting and engineering traffic are competing with each other for the same LAN resources.

Often the geography of LANs makes it impossible to physically attach multiple LANs together. The involved LANs might be located in different countries, states, cities, streets, buildings, or even floors. In these cases, LAN-to-LAN traffic must be transported over a different medium (e.g., a fiber optic link, a digital circuit, or even an analog phone line) in order to join the LANs together. The traffic that flows across the LAN-to-LAN link must be carefully regulated, because LAN-to-LAN links usually operate at lesser speeds than the LANs themselves. If all traffic is funneled through the lower-speed link, the effective speed for each LAN can be adversely affected.

Finding a practical balance between the flow of LAN-to-LAN traffic and overall network performance is the challenge of interconnecting LANs. Too much LAN-to-LAN traffic diminishes performance. Too little traffic curtails value. No single product can address the range of problems created by LAN interconnections, but bridges and routers go a long way toward solving the majority of interconnection issues.

□ Bridges—Bridges are relatively unintelligent devices that forward information from one LAN to another based on information contained in the LAN frame headers. Bridges are not fully aware of network protocols (e.g., TCP/IP, IPX, SNA, etc.)—most bridges can't even distinguish between an Ethernet frame carrying TCP/IP and an Ethernet frame carrying IPX.

□ Routers—Routers represent an evolutionary step beyond bridges. Like bridges, routers are devices that forward information between LANs, but unlike bridges, routers determine what traffic to "route" based on information maintained at the network protocol level (e.g., TCP/IP, IPX, SNA, etc.). For example, in a combined TCP/IP, IPX, and SNA network, a router can be configured to forward only selected IPX traffic between LANs.

A bridge and router are sometimes combined to create a third type of device—a "brouter." A brouter is a hybrid device that provides both bridging and routing services.

As shown in Fig. 4-1, bridges and routers can be employed as single units or used in pairs. When used alone, a bridge or router connects to two or more LANs using native LAN connections (e.g., Ethernet or Token Ring). This technique is used to interconnect LANs that are physically close to one another. For example, a bridge or router can be attached to two Ethernet LANs and forward selected traffic between the two.

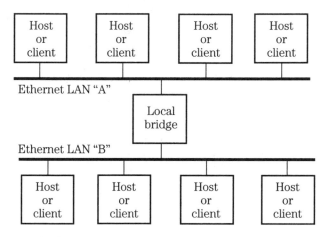

A local bridge connects two (or more) physically adjoining LANs.

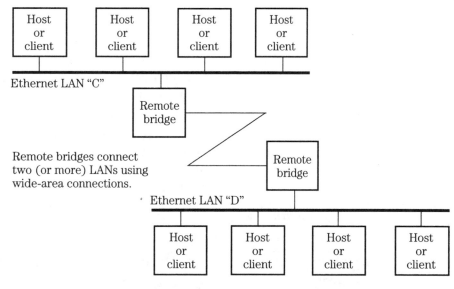

■ **4-1** *Local versus remote bridges*

More commonly, bridges and routers are used in conjunction with fiber optic or other wide area links to interconnect LANs located far from one another. When this technique is used, two or more bridges/routers are required—one bridge/router takes selected traffic from the LAN and forwards it over the wide area link to a second bridge/router. The second bridge/router then receives the traffic from the wide area link and places it on the LAN it is connected to.

Bridges and routers also provide the key capability of interconnecting different types of LANs. For example, a bridge may be used to join a local Ethernet network with a local Token Ring network. Similarly, a pair of routers may be used to move traffic between a local Token Ring network and a remote Ethernet network. In short, bridges and routers allow logical network interconnections to transcend the issues of conflicting physical networks.

Access protocols & network protocols

Bridges, routers, and brouters come in all sizes and shapes, offer various levels of security and manageability, and provide a variety of functions and services. Before we can delve further into these subjects, however, we must explore the fundamental difference between bridging and routing—that is, how the LAN access protocol differs from the network protocol.

Every LAN operates a low-level protocol that defines the rules for sending and receiving messages over the physical LAN medium. This protocol is referred to as the "access protocol," because it regulates how devices get information into and out of a LAN. The access protocol must be observed by all devices participating in a LAN—it is the common denominator for communications.

The access protocol is unique to each type of LAN and defines the access method (e.g., token-passing or carrier detection) as well as the data structure (the frame) used to get information into and out of the LAN. The frame is particularly important because it defines the address of the originating system (the source address), the address of the target system (the destination address), and other control-oriented information.

As discussed in Part II, the modern world of LANs is dominated by these four LAN types:

☐ IBM Token Ring—Although the design for a Token Ring network dates back to 1969, IBM's formal involvement with Token Ring did not occur until the early 1980s. In 1982, IBM presented a series of papers on Token Ring technology to the Institute of Electrical and Electronics Engineers (IEEE). While the IEEE mulled over these papers, IBM went on to produce its own commercial version of a Token Ring LAN, released in 1985.

☐ IEEE 802.5 Token Ring—The IEEE took IBM's 1982 recommendations, made a few minor adjustments here and there, and created the 802.5 standard for a Token Ring LAN. These specifications were released in 1985.

☐ IEEE 802.3—Prior to their work on the 802.5 specifications, the IEEE took the specifications for Ethernet Version 1.0 (as defined by Xerox, Digital Equipment Corporation, and Intel Corporation in 1980) and created the specifications for a Carrier Sensing, Multiple Access with Collision Detection (CSMA/CD) LAN. These specifications, known as the 802.3 standard, were released in 1983.

☐ Ethernet Version 2.0—The IEEE 802.3 specifications differed from Ethernet Version 1.0 to the extent that the two LAN types were incompatible with one another. In response, Xerox, Intel, and Digital Equipment created the Version 2 implementation of Ethernet so that 802.3 and Ethernet could coexist on the same physical LAN. Version 2.0 was released in 1982.

These four types of LANs can be roughly divided into three categories of access protocols: token-passing, IEEE 802.3 CSMA/CD, and Ethernet CSMA/CD. In this breakdown, both IBM and IEEE 802.5 use the same token-passing methodology (even though their frame structures are not identical). The two CSMA/CD implementations—IEEE 802.3 and Ethernet—actually share the CSMA/CD access methodology, but their frame structures are different to the extent that they should be regarded as different access protocols.

Also note that other types of LANs are in use in the marketplace—in particular the Fiber Distributed Data Interface (FDDI) on the high end and ARCnet on the low end—but these LANs do not enjoy any significant market share in contrast to the big four.

Beneath the access protocol is yet another set of protocols called "network protocols." These protocols are implemented as part of a larger network architecture (such as IBM's SNA, Digital's DECnet, Novell's NetWare, or TCP/IP) and are carried within the "information" field of the access protocol frame, as shown in Fig. 4-2. This means that as far as the access protocol is concerned, all network protocols look the same, because they are all simply carried as data in an access protocol frame.

Start frame delimiter	Access control	Frame control	Destination address	Source address	Source route (optional)	Destination Service Access Point (DSAP)	Source Service Access Point (SSAP)	Control	Information (data)	Frame check sequence	End delimiter	Frame status

Access protocol frame (Token Ring)

Network protocol frame (IPX)

Checksum	Packet length	Transport control	Packet type	Destination network	Destination node	Destination socket	Source network	Source node	Source socket	Information (data)

■ **4-2** *Net protocols within access protocols*

82

Although LAN access protocols don't care about network protocols, sending and receiving systems certainly do. For example, when a system receives a message, it must look at the information field in the access protocol frame and figure out what network protocol is being used. The receiving system can then take appropriate action based on the information carried in the network protocol data structure.

This separation between the access protocol and the network protocol is the mechanism that allows a single LAN (a single access protocol) to carry multiple network protocols. Does this mean that each system on a LAN must be aware of all of the network protocols operating beneath the access protocol?

Not at all. In fact, a system supporting only one network protocol will often never see traffic generated by other network protocols, because a system using one network protocol does not typically send messages to other systems that don't use that same network protocol. For example, IBM SNA systems aren't inclined to start conversations with Novell NetWare servers. This means that although the physical network may be teaming with different network protocols, each system only "sees" the network protocol(s) it is concerned about.

One exception to this "to each his own" rule is that some systems send "broadcast" messages, which are, in fact, delivered to every system in the network. For example, Novell NetWare servers frequently broadcast their addresses and the services they offer so NetWare clients can "find" them. Fortunately, a system sending a broadcast message does not expect every system in the network to respond to it, so those systems that don't participate in that network protocol can simply ignore the broadcast message.

Thus far, this discussion has assumed that each system is oriented toward a specific network protocol (or suite of related protocols), but in reality this is becoming less and less true. For example, a few years ago a PC might understand a single network protocol like Novell NetWare or Banyan VINES, but today that same PC might participate in several different network protocols—such as Novell NetWare, TCP/IP, and IBM SNA—all at the same time.

Participation in multiple protocols is even more common at the server level—today's LAN servers don't want to be pigeonholed into a particular network protocol, they want to provide a variety of services to a wide range of clients using whatever protocols are necessary or appropriate. For example, in the not-too-distant past, a LAN might support a Novell NetWare server providing file services to PC clients via the IPX network protocol, and a Unix NFS server providing file services to Unix workstations via the TCP/IP protocol suite. Today, however, a single multipurpose, multiprotocol "superserver" can provide file services to both types of clients using both types of protocols.

How do systems participating in multiple network protocols figure out which message pertains to which protocol? And what does this have to do with bridges and routers? Well, as it turns out, routers, servers, clients, and other LAN-attached devices all use the same mechanisms to determine which network protocol is being carried in an access protocol frame.

Network protocol identification

As discussed, a single LAN access protocol can sponsor many different network protocols because the encoding specifications are well-defined, well-understood, and (most importantly) adhered to by network developers. To a certain extent, conformance to these specifications is self-enforced, because the risk of deviating from the specification is the risk of compromising the integrity and functionality of a network—and no company wants that kind of exposure. Therefore, in the world of LANs, Microsoft NetBIOS works alongside Novell IPX, IBM SNA coexists with Digital DECnet, and TCP/IP runs next to Banyan VINES.

All access protocols provide a relatively simple means of determining what network protocol is being carried in the frame. The exact methodology, however, is dependent on the specific access protocol.

For IEEE 802.3 (CSMA/CD), IEEE 802.5 (Token Ring), and IBM's Token Ring implementation, the network protocol is initially determined by examining the values contained in the Destination Service Access Point (DSAP) and Source Service Access Point (SSAP) fields. The two SAP fields serve to identify the network protocol that originated the message (SSAP) and the network protocol where the message is to be delivered (DSAP). Normally both the SSAP and DSAP contain the same value—messages do not usually bounce between network protocols. SAP values are assigned by the IEEE. A list of commonly encountered assignments is shown in Table 4-1.

■ Table 4-1 IEEE 802.2 SAP assignments.

SAP	Assignment
04–05	IBM SNA
06	ARPANET Internet Protocol (IP)
18	Texas Instruments
7E	X.25 Level 3
80	Xerox Networking Services (XNS)
98	Address Resolution Protocol (ARP)
BC	Banyan VINES
E0	Novell NetWare
F0	IBM NetBIOS
F4–F5	IBM LAN management
F8	IBM Remote Program Load (RPL)
FA	Ungermann–Bass
FE	Open System Interconnect (OSI) network layer

For Ethernet Version 2.0 networks, a single Protocol Type field is used to identify the protocol. As in the case of the 802 SAPs, a variety of Ethernet types have been predefined for specific protocols and/or specific vendors. These assignments are shown in Table 4-2.

■ Table 4-2 Ethernet type assignments.

Type	Assignment
0600	Xerox Networks Services (XNS)
0800	ARPANET Internet Protocol (IP)
0801	X.75 Internet Protocol (IP)
0802	NBS Internet Protocol (IP)
0803	ECMA Internet Protocol (IP)
0804	CHAOSnet
0805	X.25 Level 3
0806	Address Resolution Protocol (ARP)
0BAD	Banyan VINES
6000–6009	Digital Equipment Corporation
6010–6014	3Com Corporation
7000–7002	Ungermann–Bass
7030	Proteon
7034	Cabletron
8005	Hewlett-Packard probe protocol
8010	Exelan
8013–8016	Silicon Graphics
8035	Reverse ARP
8038–8042	Digital Equipment Corporation
8046–8047	AT&T
807D–8080	Vitalink
8088–808A	Xyplex
809B	Ethertalk
80C0-80C3	Digital Communication Associates (DCA)
80D5	IBM SNA
80F2	Retix
80F3–80F5	Kinetics
80F7	Apollo Computer (HP)
80FF–8103	Wellfleet Communications
8137–8138	Novell

In all types of LANs, the process of determining a network protocol begins when a system receives either a broadcast frame or a frame specifically addressed to it. After receiving the frame, the low-level network software looks at the SAP or Protocol Type field to ascertain whether it can understand the network protocol involved. If the system does not recognize the network protocol, it simply ignores the message.

Network protocols

Now that we have a well-understood mechanism to identify the network protocol carried within a LAN frame, we know everything we need to know about any given LAN message, right?

No, not really. In order to efficiently handle a variety of functions and applications, most network protocols are subdivided into services, and each service has its own unique data format and handshaking requirements—in effect, the network protocol is broken down into service protocols, as shown in Fig. 4-3. We can see this clearly if we take a close look at some popular network protocols.

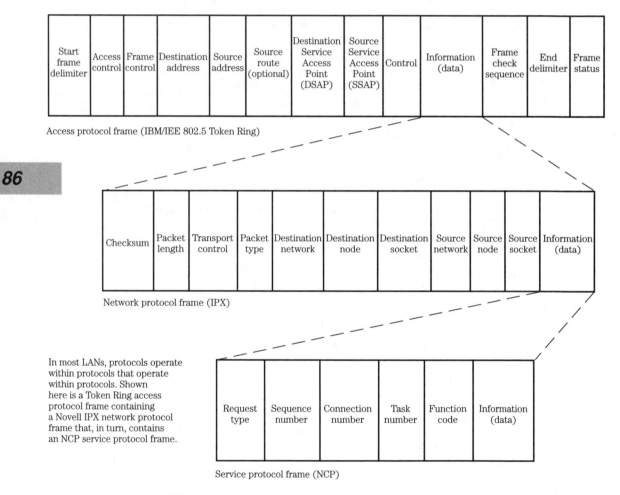

Access protocol frame (IBM/IEE 802.5 Token Ring)

Network protocol frame (IPX)

In most LANs, protocols operate within protocols that operate within protocols. Shown here is a Token Ring access protocol frame containing a Novell IPX network protocol frame that, in turn, contains an NCP service protocol frame.

Service protocol frame (NCP)

■ **4-3** *Access/network/service protocol layers*

Novell NetWare

The network protocol used in the Novell NetWare environment is the Internetwork Packet eXchange (IPX). IPX is a multipurpose transport that can carry a number of service protocols, including the Sequenced Packet eXchange (SPX) protocol. By itself, IPX is a "connectionless" protocol that does not guarantee delivery of messages. SPX, on the other hand, is a "connection oriented" protocol that runs as an extension to IPX and provides confirmation (or denial) of the end-to-end delivery of messages.

NetWare service protocols can run under just IPX, or the IPX/SPX combination. These services include

- [ ] NetWare Core Protocol (NCP)—This protocol handles the mainstream NetWare services, including accessing files and printers on NetWare servers.
- [ ] Burst Mode protocol—This is a variation of the NetWare Core Protocol. Burst Mode allows a client to request and receive more data in a single message than under NCP. Burst Mode was designed for high-volume applications.
- [ ] Service Advertising Protocol (SAP)—File, print, communication, and other types of servers announce themselves at regular intervals using this protocol. Client PCs "listen" for this protocol to determine what resources are available within the network. Clients can also use this protocol to inquire about the capabilities of specific servers.
  Note: The Novell NetWare Service Advertising Protocol (SAP) is an entirely separate entity from the IEEE Service Access Point (SAP).
- [ ] Routing Information Protocol (RIP)—This protocol is used to help a message move from one NetWare network to a second NetWare network. As we will see later in this chapter, routing protocols like RIP are an important factor in how routers operate.

TCP/IP

Like IPX, the Transmission Control Protocol and Internet Protocol (TCP/IP) are multipurpose transports. TCP/IP is, however, somewhat different from other network protocols because it is, in fact, two layers of protocols—TCP and IP—and each layer can transport additional protocols. For example, IP can transport

- [ ] Transmission Control Protocol (TCP)—TCP is a connection-oriented protocol that guarantees end-to-end delivery. TCP is

87

used as the primary transport for the majority of the TCP/IP service protocols.

□ User Datagram Protocol (UDP)—UDP is a connectionless protocol that does not guarantee end-to-end delivery. Because it is connectionless, UDP is faster than TCP. UDP is typically used for real-time (read "client/server") program-to-program communication that requires the fastest possible response time.

As noted, TCP supports dozens of different service protocols. Some of the more notable and popular services include

□ Telnet—Telnet provides a means for a TCP/IP workstation or a terminal attached to a TCP/IP host to access a second host system.

□ File Transfer Protocol (FTP)—FTP facilitates the movement of text and binary files between systems.

□ Simple Mail Transfer Protocol (SMTP)—SMTP handles the routing of mail in a TCP/IP network.

□ Simple Network Management Protocol (SNMP)—SNMP provides the framework for systems to report problems, configuration information, and performance data to a central network management location.

□ Routing Information Protocol (RIP)—This protocol is similar to the Novell's RIP (they have a common parent), and is one of several routing protocols that can be used in conjunction with TCP/IP. We'll come back to the subject of routing protocols in a little bit.

Many more service protocols are run under the umbrella of TCP/IP. This large number and wide range of available service protocols is another aspect that separates TCP/IP from other network protocols—no other network protocol has so many optional services.

IBM SNA

IBM's System Network Architecture (SNA) is really an entire family of protocols based on different Physical Unit (PU) and Logical Unit (LU) classifications. Mainframes use the entire range of IBM SNA services, while IBM AS/400 systems predominantly depend on the PU 2.1 (with or without the Advanced Peer-to-Peer (APPN) extensions) and LU 6.2 (a.k.a., APPC) combination for peer-to-peer communication and for PC communication.

As was the case for TCP/IP, a number of service protocols ride on the coattails of specific PU/LU combinations. For example, the combination of PU 2.1 and LU 6.2 can carry

- [ ] Distributed Data Manager (DDM)—DDM allows access to physical or logical files on remote AS/400 systems.
- [ ] Display Station Passthrough (DSPT)—DSPT lets a workstation attached to one AS/400 access a second AS/400 as if it were a native workstation.
- [ ] SNA Distribution Services (SNADS)—SNADS is an object distribution service that can be used to transport mail, documents, files, and other objects over the network.
- [ ] Advanced Program-to-Program Communication (APPC)—APPC is synonymous with LU 6.2. APPC provides an interface that programs can use to communicate with one another over the network.

Unlike IPX and TCP/IP, SNA does not have service protocols dedicated to the task of routing. SNA depends on access protocol services such as the "source routing" of Token Ring LANs to perform these functions. For more information about SNA as it relates to internetworking, please refer to the heading *Routing and SNA* later in this chapter.

Bridging

Now that we have a better appreciation of the two major levels of protocols in a LAN—the access protocol and the network protocol—we can nail down the differences between bridges and routers. Of the two, bridges are the easiest to understand because they operate at the simpler access protocol level. Let's explore these two devices, starting with bridges.

In the simplest terms, a bridge takes a frame off the local network and sends it over a wide area link to another bridge where it is inserted onto another LAN. This flow is shown in Fig. 4-4. Bridges of varying intelligence are available, and this intelligence is used to determine when it is appropriate to forward a frame over the wide area network.

At the simplest (least intelligent) level, bridges send every single frame across the wide area link, without regard to where the receiving system actually resides. The disadvantage of this approach is that it creates a fair amount of extraneous traffic on the remote network. The advantage is that it is simple to implement and it is oblivious to changes in the network.

Since simple bridges cause superfluous traffic to travel over the wide area link and onto both LANs, more intelligent bridges have been developed to send selected frames across the wide area link.

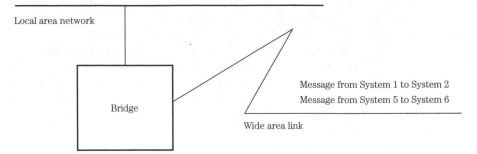

Message from System 1 to System 2 | Message from System 3 to System 4
Message from System 5 to System 6 | Message from System 7 to System 8

Local area network

Bridge

Message from System 1 to System 2
Message from System 5 to System 6

Wide area link

A bridge forwards information from one LAN to another based on simple filtering criteria.

■ **4-4** *Local bridge message flow*

Two different techniques are used to instill address intelligence into this type of bridge:

☐ A manually configured bridge is "told" what destination addresses are on the other side of the link. These addresses can be configured as a list of selected addresses, as a series of address ranges, or even as an address mask. Once the remote addresses are established, the bridge only sends frames across the wide area link that contain one of the selected addresses in the destination address field. This approach minimizes traffic over the wide area link (and on the remote LAN), but implementing it requires some advance planning when it comes to assigning network addresses.

☐ A self-learning bridge "observes" which addresses are in which network and then only sends frames across the wide area link for the remote addresses. During the learning phase, this type of bridge sends all information across the bridge, and observes where the response actually comes from. The advantage of this type of bridge over a manually configured bridge is that a learning bridge does not require any reconfiguration when systems are added (or deleted) from any of the connected LANs.

Also note that some bridges can filter information based on the Service Access Point or Protocol Type fields. This type of filtering, however, puts bridges in the same neighborhood where low-end routers live.

Routing

The inadequate performance of the first generation of bridges combined with the proliferation of network protocols to create a real need for better LAN-to-LAN traffic management. Routers were invented to address this need by managing LAN-to-LAN traffic flow based on information maintained at the network protocol level. In large networks running multiple network protocols, routers offered the perfect solution—those protocols that needed to travel over wide area links could do so, and those that had nowhere to go did not make the trip.

Routers cannot, however, act as universal replacements for bridges because of one very important fact: although all network protocols can be bridged, they cannot all be routed. The key to understanding which protocols are routable and which are not lies in how a network protocol handles the addressing of messages.

Let's assume that we establish a rule that says each separate LAN must have a unique network address. If a network protocol uses an addressing scheme that includes network addresses (in addition to individual system addresses), then a network-level device (e.g., a router) can easily detect when a message needs to move between networks, because the source network address and destination network address are not the same.

For example, if system "1" in network "0" wants to send a message to system "10" in network "2," it includes its local network ("0") and system ("1") identifications in a source address field, and the target network ("2") and system ("10") identifications in a destination address field. When a router attached to network "0" gets the message, it looks at the source and destination address and can "see" that the message needs to be forwarded to network "2." This flow is shown in Fig. 4-5.

Where are these address fields kept? That depends on the network protocol. Both IPX and TCP/IP have separate fields within their own data structures to define source and destination addresses that include both network and system identifications. This means that when either IPX or TCP/IP is being used, a LAN frame has two sets of addresses—the source and destination addresses maintained within the access protocol data structure, and the source and destination address maintained in the network protocol.

Does the address information in the network protocol override the access protocol addresses? No. Access protocol addresses are required by the access protocol to deliver a message to a specific sys-

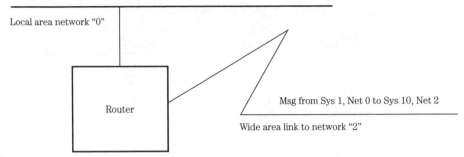

Msg from Sys 1, Net 0 to Sys 10, Net 2 | Msg from Sys 2, Net 0 to Sys 4, Net 0

Msg from Sys 5, Net 0 to Sys 22, Net 7 | Msg from Sys 7, Net 0 to Sys 8, Net 0

Local area network "0"

Router

Msg from Sys 1, Net 0 to Sys 10, Net 2

Wide area link to network "2"

■ **4-5** *Message flow through a router*

tem. So how do the two relate? In simple terms, the network proto-col address provides the "big picture" address—that is, the location of a system within an entire, enterprise-wide network—while the access protocol details the location of a system within a LAN.

The two types of addresses work together, although the exact methodology varies from one network protocol to another. In the Novell NetWare environment, the access protocol address for the server is included in Service Advertising Protocol (SAP) messages, which are regularly broadcast to all clients. The client can use that address to send messages directly to the server. Similarly, when the client initiates communication with a server, it includes its own access protocol address, so the server knows how to directly address it.

In the TCP/IP environment, the Address Resolution Protocol (ARP) is used to convert network protocol addresses into access protocol addresses and vice versa. ARP is a service protocol that operates under IP.

Routing & SNA

Not all network protocols use a second set of addresses. For example, in a LAN environment, SNA uses access protocol addresses for communication. Since these addresses tend to be irregular and somewhat random, most implementations of SNA software provide mechanisms to override the hardware address with a more sensible address so you can establish some sort of address scheme. For instance, the IBM LAN Support Program (LSP) drivers (e.g., DXMC0MOD.SYS and DXME0MOD.SYS) allow PC LAN adapter addresses to be redefined.

So is the absence of a second set of addresses in SNA the reason why SNA is not a routable protocol? That is, in fact, correct. But does that mean that SNA must always be bridged and never routed? Well, not exactly. SNA has two provisions for routing:

☐ IBM Token Ring LANs include support for a feature called "source routing." This is kind of a mini-routing protocol that is handled at the bridge level (i.e., at the access protocol level). IBM is the only major network vendor that implements routing in bridges instead of routers. See Chapter 3 for more information about source routing.

☐ Under SNA (and APPN), routing decisions can be made within IBM host computers. For example, if you implement an APPN network with AS/400s, the AS/400s act as the routers. The design of SNA (and APPN) does not easily accommodate the off-loading of the routing decision-making process to a separate box.

If you're really bent on routing SNA using a third-party router, you do have alternatives for handling SNA traffic.

The first option is to use routers that have the capability to forward all of the traffic for a specific protocol (or protocols). Operating in this mode, a router looks at the appropriate field in the access protocol data structure (e.g., the Destination Service Access Point or Protocol Type field), determines what the network protocol is, and if it is a selected protocol, forwards it to another network without further inspection or evaluation.

Routing an entire network protocol is really a crude solution because no effort is made to filter traffic—in effect, the router becomes a bridge for one or more specific network protocol(s). This is also a gray area where low-end routing runs into high-end bridging.

The second option is to use routers that support Data Link Switching (DLSw). DLSw is a relatively new routing protocol designed to "switch" SNA traffic through a TCP/IP network (the designers of DLSw prefer to use the term "switch" instead of "route").

In a LAN environment, DLSw routers listen for the "test" message that a system uses to verify that another system is present before it transmits to it. When a DLSw router hears a test message, it sends an inquiry to other DLSw routers to see whether the requested system is remotely located. If the system is remote, the

DLSw router responds to the test message, telling the originating system that the target system is local (even though it really isn't).

When the source system starts transmitting to the target system, the DLSw router takes the SNA information, encapsulates it in TCP/IP frames, and sends it to the next involved DLSw router. Under certain circumstances, DLSw routers will respond to the originating system instead of waiting for the response to come from the target system—this approach avoids lengthy delays caused by trying to perform polling through routers.

The specifications for DLSw have undergone several changes since the first series of DLSw routers were released. This means that although a number of DLSw routers are available on the market, many of them have subtle differences from one another. The best advice here is simple: read the fine print.

Routing protocols

For those network protocols and architectures that do support network-level addresses, the job of a router is to make sure that messages addressed to one network get delivered there. The difficulty of that job is directly related to the overall size of the network.

For example, if you only connect two LANs together, it's pretty easy for a router to know where to send a message for a different network. If, however, you string three or more LANs together, it certainly becomes less obvious, because no single router is visible to all of the networks. Now imagine that you are interconnecting a dozen networks—this means that each router only knows a piece of the network puzzle.

If each router only knows about a portion of the network, how can you get a message from one end of the network to the other? Simple—the routers have to share their knowledge with one another—they have to communicate with one another. Enter the routing information protocol.

The primary purpose of a routing information protocol is for routers to tell other routers about the network connections they maintain. Each router collects all of the information it receives from the other routers and builds its own routing table. This table determines how to move traffic for the local systems it is handling to other locations in the network.

Routers communicate with one another quite frequently. When a router comes onto the network, it requests information from other routers. When a router is about to go off the network (orderly shutdown), it tells the other routers so that they can build alternate routes. Some routing protocols even have routers broadcast their routing information on a regular basis (e.g., every 60 seconds) to make sure that the information stays current for all routers.

Client and server systems also get involved with routing protocols. When a client system wants to connect to a server in another network, it broadcasts a routing protocol request to locate that server. If the router already knows how to reach the server (based on its routing table), it tells the client that everything is okay. Otherwise, the router broadcasts a request to the other routers to find a path to the requested server. Once the path is discovered, the involved routers update their routing tables, and the client is given a go-ahead to proceed.

Many different implementations of routing information protocols are used in modern networks. Some vendors supply one or more with their proprietary network implementations, and several routing protocols are defined to address routing in "open" networks like Open System Interconnect (OSI) and TCP/IP. The routing protocols you'll most likely encounter are

☐ Routing Information Protocol (RIP)—RIP was developed to provide routing services in Xerox Network System (XNS) networks. The XNS implementation was subsequently adapted for use in the Novell NetWare environment and in the TCP/IP environment. Routers using RIP broadcast their routing tables to one another on a regular basis, as shown in Fig. 4-6.

One aspect of RIP that has come under fire recently is that RIP assumes a path that passes through the fewest routers (has the fewest "hops") is the best path to take. Unfortunately, that is not always the case, because different routers may be connected using different-speed links. Under RIP, for example, a path composed of two 9600 links is considered better than a path composed of three 64,000 bps links, because the first path has fewer hops (two instead of three).

☐ Open Shortest Path First (OSPF)—The OSPF routing protocol was developed to replace RIP in the TCP/IP environment. OSPF has a number of advantages over RIP:

~ OSPF operates at the IP level within TCP, so it is capable of routing a wider variety of services than RIP, which operates at the TCP level.

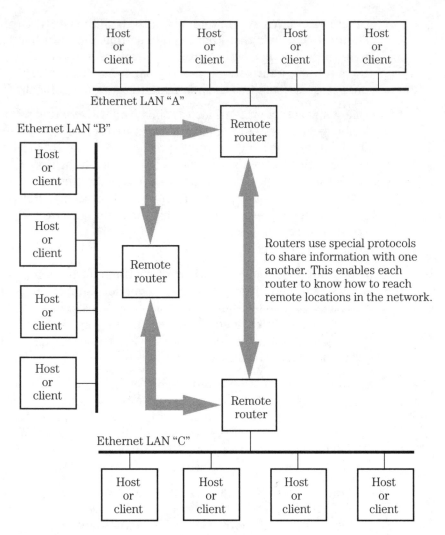

Ethernet LAN "A"

Ethernet LAN "B"

Routers use special protocols to share information with one another. This enables each router to know how to reach remote locations in the network.

Ethernet LAN "C"

■ **4-6** *Router table broadcasts*

~ OSPF is not as "boisterous" as RIP. Whereas RIP will broadcast new routing tables to all routers on the network at the drop of a hat, OSPF dictates that routing table updates get distributed to neighboring routers only. This results in less administrative traffic in the network without compromising routing operations.

~ OSPF takes into account a number of factors to determine the best possible route—not just the number of "hops." In particular, OSPF takes into consideration issues like line speeds, error rates, and current loading to arrive at the best path for a message.

Note also that other routing protocols are used in TCP/IP networks by TCP/IP gateways. These routing protocols will be discussed in Chapter 5.

Does the availability of so many routing protocols complicate LAN-to-LAN networking? Yes it does. But implementing and managing a large, multi-LAN network is pretty complicated at all levels.

Brouting

Given that both bridges and routers have their place in wide area network society, many manufacturers have combined the two technologies into a single unit, often called a "brouter." As the name implies, a brouter is capable of performing both bridging and routing functions.

Like a router, a brouter operates at the network protocol level. At that level, a brouter is capable of routing protocols in the same fashion as a standard router. However, a brouter may also be configured to bridge protocols—that is, whenever a brouter detects specific vendor protocols (such as LAT or NetBIOS), it performs bridging actions at the physical protocol level.

In real-life terms, the brouter is a junction point for bridges and routers, and most manufacturers are headed in that direction. Routing is being introduced into bridges and bridging is being introduced into routers. Although not everyone calls the resulting device a "brouter," this technology and its advantages remain the same—a brouter by any other name performs the same function.

97

Other considerations

Given the differences between routers and bridges discussed in this chapter, the question becomes which is appropriate for your network? At one level, this is an easy question to answer, because if your network protocols do not use network addresses, they must be bridged, and if they do use network addresses, they can be routed. Please note, however, that the characteristics of the access protocols also have a bearing on your choice.

IBM Token Ring LANs, as well as other LANs that follow the IEEE 802.5 networking standard, have a feature that further confuses and complicates the issues of bridges and routers. This feature is known as "source routing." This feature is discussed in detail in Chapter 3, but a brief review will show how it affects the deployment of bridges and routers.

With source routing, the system sending information is responsible for determining where the destination address is. When a sending station recognizes that the intended receiver is not in the local network, it sends out a special message, called a "route discovery" message. The purpose of this discovery message is to find out where the receiving station is in the wide area network, and what the best possible path is to send it information.

Bridges in the LAN see this discovery message and send it over the wide area links. If additional bridges are on the remote LANs, they also send the discovery message on to other remote systems. In short order, this discovery message is propagated through the wide area network.

When the receiving station responds to the discovery message, the routing information is returned to the bridges and to the initiating station. At that point the initiating station logs the routing information and then includes it as a destination address (using a special format) in the physical protocol frame. When a bridge sees this special address, it uses the embedded routing information to deliver the data.

Source routing is implemented at the access protocol level and not the network level, so source routing is only implemented in bridges and brouters. For this reason, you don't find many traditional routers in Token Ring environments that use source routing (as in the case of most IBM networks).

Because source routing is an optional feature of IEEE 802.5 LANs, some vendors choose to ignore it in their access protocol software. This is not a problem when all of the systems in a network do not support source routing—they just use traditional (nonsource routing, or "transparent") bridges and routers. On the other hand, when systems using source routing are intermixed with systems not using source routing, the network becomes much more complicated.

In a mixed environment, systems using source routing cannot direct traffic over routers or nonsource routing bridges, and similarly, nonsource routing systems cannot forward traffic through source routing bridges. To solve this problem you must either implement a source routing bridge and a second nonsource routing bridge or router, or implement a bridge or brouter that supports both source routing and nonsource routing traffic (this is sometimes called "Source Route Transparent" (SRT) bridging). Either way you look at it, this is a more expensive solution than standardizing on one type of traffic.

Although Ethernet and IEEE 802.3 LANs do not have any technical quirks like source routing, they do share the rich history of Ethernet networking. Ethernet has been in use for a long period of time, and it has seen the comings and goings of many different protocols, bridges, routers, and other assorted LAN oddities.

When bridges were first introduced to the Ethernet environment, they were of the least intelligent breed, and routed all traffic over the wide area link. As previously noted, this created a great deal of extraneous traffic on both LANs, and had the side effect of giving bridges a bad reputation in the Ethernet community. This, of course, led to the development and deployment of routers and more intelligent bridges.

As it stands today, many Ethernet and 802.3 networks utilize both bridges and routers. One reason for using both types of devices is the fact that some network protocols must be bridged while others can be routed. But another reason is that bridges have always been an important component in building Ethernet LANs—especially bridges that directly interconnect two or more physically adjacent LANs.

Gateways & hubs

GATEWAYS AND HUBS ARE HIGHLY SPECIALIZED DEVICES that can be employed to interconnect networks. The formal definitions for gateways and hubs are somewhat sketchy. Gateways, for example, can be composed of hardware, software, or combinations of both, and provide functions ranging from simple protocol encapsulation to sophisticated protocol conversions. Hubs are clearly hardware products, but they too offer a range of services, starting with basic wiring management and extending upward into the network management, bridging, routing, and even gateway functions.

Both devices can be deployed to interconnect multiple LANs (or LAN segments), multiple wide area networks, or combinations of the two types of networks. This flexibility separates them from connectivity devices such as bridges and routers, because bridges and routers are focused on interconnecting two or more LANs.

From the broadest possible perspective, the basic functions of gateways and hubs can be described as follows:

A *gateway* provides a means of interconnecting dissimilar network types or network protocols. In some cases, the interconnection provides shared access to a common resource, and in other cases the connection involves sophisticated translation of network protocols and services. For example, a gateway can facilitate 3270 workstation access for multiple PCs on a Novell LAN by routing the PC traffic through a single gateway device that attaches to the mainframe via a Synchronous Data Link Control (SDLC) line. Alternatively, a gateway can provide interoperability services such as workstation access, file transfer, and program-to-program communications among computers on an IBM SNA Token Ring LAN and computers on a Digital DECnet LAN.

A *hub* provides a central location for the joining of multiple network attachments. These attachments might be LAN segments, wide area links, fiber connections, unshielded twisted pair (UTP) attachments, shielded twisted pair (STP) attachments, or combinations of these different technologies. Once connected, the traffic from these network links is rerouted (and possibly manipulated) by the hub. For example, a hub can provide multiple UTP connections to workstations and then route that traffic over a single Ethernet LAN attachment. Or a hub can combine traffic from an Ethernet LAN, a Token Ring LAN, and even an SDLC connection and route it over an FDDI link.

As you can see, both of these devices can be used for a number of applications. These applications, and the technology behind gateways and hubs, will be further explored in this chapter.

Gateways

The dictionary describes a gateway as a "means of access." Although this definition clearly embraces the spirit of network gateways, a more technical description for a gateway is "a device or point of presence that interconnects dissimilar networks." In this enlightened definition, the key concept is "dissimilar networks," because networks can be dissimilar in a number of ways. For example, Ethernet and Token Ring LANs are dissimilar. IBM Advanced Peer-to-Peer Networking (APPN) and TCP/IP networks are dissimilar. Wide area SDLC and metropolitan area FDDI networks are dissimilar. In truth, it is usually easier to find dissimilarities between networks than it is to find similarities.

Given this range of disparity, the question becomes "What kind of dissimilarities will gateways attempt to address?"

Interestingly enough, gateways attempt to resolve most dissimilarities between networks, although rarely does a single gateway attempt to do it all. This may lead to deploying different gateways for different problems. For example, one gateway might be put in place to provide common access to a wide area link from a local area network, while a second gateway might be installed to translate IBM SNA services into TCP/IP services.

The environments that might benefit from the installation of a gateway are numerous—too numerous to address on an individual basis. Instead we can look at gateways from a functional perspective and separate them into the following logical categories:

☐ Point-of-access—Gateways in this category manage multiple, possibly concurrent access to a common network resource. For example, a "modem gateway" can be installed on a PC LAN that allows PCs to contend for access to dial-out services.

Or a PC on a LAN can be dedicated as a "remote control" gateway to allow remote PCs to dial up and access all of the services of the LAN.

☐ Transport—This kind of gateway carries data from one type of network over a second type of network. This is typically handled by embedding all of the information associated with one protocol inside another protocol. Two scenarios exemplify this approach:

~ Traffic from one type of network can be routed through a different type of network. For example, IBM SNA traffic can be routed through a TCP/IP network if one gateway inserts SNA protocol information from one SNA subnetwork into a TCP/IP network and another gateway extracts the SNA data from the TCP/IP network at a different geographical location and forwards it to another SNA subnetwork. This is sometimes referred to as *tunneling*.

~ To reduce the number of protocols required on a LAN, client-side software can embed foreign protocol information in the local protocol and route the traffic to a gateway, which can then extract the foreign protocol and forward it to the foreign network. For example, PC background software can embed Digital Local Area Transport (LAT) information into Novell Internet Packet eXchange (IPX) frames and route them to an IPX/LAT gateway. The gateway removes the LAT information and forwards it over a separate link to an LAT-oriented device (such as a Digital host computer). This technique is often called *encapsulation*.

☐ Intranetwork—Gateways of this class serve to make connections in the context of a single network architecture (e.g., SNA, TCP/IP, NetWare, DECnet). These connections can then enable traffic flow between clients and servers operating in multiple logical networks and/or multiple physical networks (such as Ethernet and Token Ring). One of the confusing aspects of this category is that it pushes into the turf ruled by bridges and routers.

☐ Interoperability—Gateways in this category translate network and application-oriented services between two (or more) types of networks to accommodate interoperability. These types of gateways require a high degree of sophistication because they must be able to translate between network protocols, network services, and data formats. For example, a gateway that accommodates electronic mail exchange between SNA and DECnet must be fluent in the mail protocols and formats used in both network environments. Similarly, a gateway for file transfer and workstation access must translate between network file access services and also provide workstation emulation.

As you can see from this list, gateways can be separated into groups that offer dramatically different functions—an interoperability gateway, for example, is similar to a transport gateway in name only. With this in mind, it's no wonder that so much confusion creeps in when the term "gateway" is mentioned.

Point-of-access gateways

A "point-of-access" gateway allows a client system to access and interact with communication services attached to the gateway itself. In most cases, the communication between the client and the gateway will be the native network protocol (e.g., IPX, NetBIOS, etc.). Once the client attaches to the gateway, it can gain access to the services available on the "other side" of the gateway.

For example, a PC can run a program to attach to a modem gateway and dial up CompuServe, as shown in Fig. 5-1. In this case, the local network does not interact with the CompuServe network, nor does the CompuServe network interact with the local network—the gateway provides a point of access without "commingling" the two networks. Other types of point-of-access gateways include the following:

☐ A number of manufacturers produce terminal gateways for PC LANs. Normally the gateway is a separate device (often a dedicated PC) that attaches to a host network (i.e., DECnet, SNA, TCP/IP) on one side and to the PC LAN on the other side. Workstation emulation software (e.g., 3270, 5250, VT200, etc.) is then placed in the PCs, and this software uses the native LAN protocol (e.g., IPX or NetBIOS) to communicate with the gateway.

☐ A remote access gateway for PC LANs can be implemented by running "remote control" software on a LAN-attached PC that also has a modem connection. Remote PCs run partner "remote control" software that dials into the gateway PC and enables remote users to access all of the facilities of the LAN as if they were sitting at the gateway PC. In this scenario, although the remote PC communicates with the gateway PC over a standard serial line using a proprietary "remote control" protocol, the LAN only sees the gateway PC and is unaware of remote connections.

104

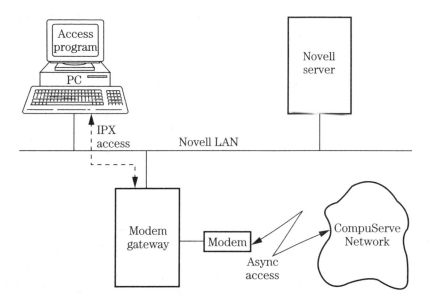

In this example, the PC user runs an access program that establishes a
connection to the modem gateway via IPX. A modem connection can
then be established to CompuServe (or any other modem-oriented
network).

■ **5-1** *Modem gateway*

Connections to point-of-access gateways are normally initiated by
the end user on a demand basis. For example, a user initiates the
connection when he or she needs to access a modem, start a mainframe terminal session, or connect to a remote LAN. Unlike most
of the other types of gateways, point-of-access gateways mask the
presence of foreign communications services from the other local
network—thus, the terminal gateway hides the host environment
from the PC LAN, and the remote control gateway hides the remote serial attachments from the local PC network.

Transport gateways

Gateways that fall into the "transport" category optimize the use
of existing network links by funneling one type of network traffic
through another network. For example, assume a large corporation has an SNA network located in Dallas, a separate (and uncon-
nected) SNA network located in New York, and a Novell IPX
network that covers the continental United States (which would of
course include Dallas and New York). If this corporation were to
deploy transport gateways, it could "attach" both SNA networks to
the national IPX network, and the gateways would route the SNA

traffic through the IPX network (invisibly to the SNA networks), thus enabling communication between the two SNA networks. This solution is shown in Fig. 5-2.

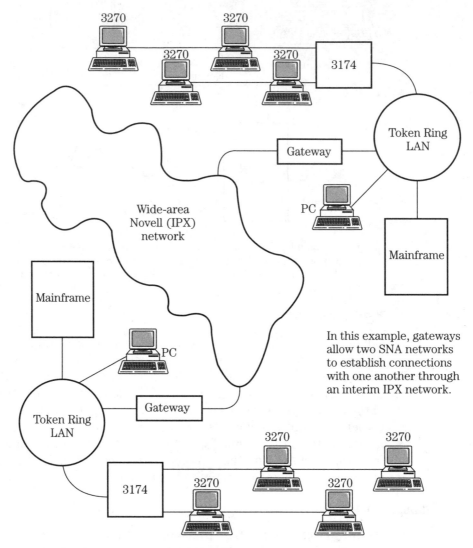

In this example, gateways allow two SNA networks to establish connections with one another through an interim IPX network.

■ **5-2** *SNA → IPX → SNA gateways*

Another use of transport gateways is to carry information from a client system over one type of network to the gateway, where it is then released into another type of network. For example, assume that a PC needs to run DECnet services, but the network it is attached to is dedicated to TCP/IP. In this case, TCP/IP can be implemented in the PC and a software module can be inserted to fool the application software into believing that DECnet is present.

As shown in Fig. 5-3, when the PC makes a DECnet request, the extra software module inserts the request into a TCP/IP frame and routes it to a gateway attached to a DECnet network. The gateway then extracts the original DECnet request, forwards it to the DECnet network, and sends the response back via the same route. As far as the PC is concerned, it is operating on a DECnet network—it does not see the TCP/IP network at all.

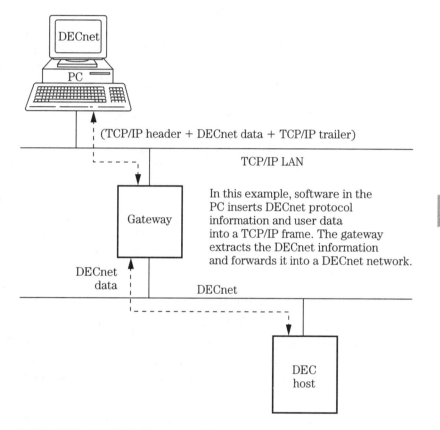

In this example, software in the PC inserts DECnet protocol information and user data into a TCP/IP frame. The gateway extracts the DECnet information and forwards it into a DECnet network.

■ 5-3 *DECnet in TCP/IP encapsulation*

Although these two examples are dramatically different, they both require two points of action. At one point, the protocol information (i.e., the protocol frame) from one type of network must be inserted into the data field of the protocol frame for the second type of network. At the other point, the original frame must be removed from the second frame and sent into the corresponding network. In the previous two examples, an SNA frame would be inserted into an IPX frame and a DECnet frame would be inserted into a TCP/IP frame.

Each of these two points of action can correspond to a dedicated hardware unit or a piece of software resident in some other system. For example, there are two gateways that use these two different approaches:

☐ The Novell NetWare for Systems Application Architecture (SAA) puts software in client PCs that allows SNA traffic (i.e., PC Support or CA/400 APPC traffic) to be carried in native Novell IPX frames. These frames are then routed to a NetWare for SAA server (the gateway), which extracts the SNA frames from the IPX frames and sends them on to an IBM host in standard SNA format. This approach is called *encapsulation*, because all client nodes use one network protocol (in this example, IPX) and foreign protocols (e.g., SNA) are encapsulated in the native protocol frames.

☐ Cisco Systems and other manufacturers produce gateways that allow SNA traffic to be routed over TCP/IP networks. In this environment, a gateway attached to both an SNA and TCP/IP network extracts frames from the SNA network, inserts them into TCP/IP frames, and sends them into the TCP/IP network. Another gateway attached to the same TCP/IP network—but to a different SNA network—extracts the SNA frames and sends them into that SNA network. This approach is called *tunneling*, because the SNA nodes have no direct participation in the TCP/IP network (they are merely passing data through it).

In both of these cases, SNA protocol frames are embedded in frames corresponding to the other type of network (IPX or TCP/IP) for transport over that network. If you were to remove a frame from the non-SNA network and look inside it, you would see that it carries an SNA frame in the data area.

Gateways that provide tunneling are invaluable if you are building complex, multivendor networks. You can, for example, link several SNA networks together using a common TCP/IP network. In contrast, gateways that provide encapsulation provide the hammer you need to force two network protocols to live together without consuming every bit of memory on your systems. A good example of this case is using NetWare for SAA to transport IBM traffic—this approach uses dramatically less PC memory than running SNA and IPX side-by-side in the PC.

Intranetwork gateways

The function of intranetwork gateways is to control the flow of traffic within a network architecture (e.g., Novell's NetWare or TCP/IP). The traditional role of an intranetwork gateway is to connect two physically separate networks. In this capacity you will no doubt recognize that the function of a gateway is extremely similar to the function of a router. For that reason many vendors have abandoned the term *gateway* in favor of *router* for this style of network-to-network connection. Please see the section entitled *Gateways versus routers* later in this chapter for more information on this subject.

One significant network architecture that continues to use the term *gateway* to describe intranetwork connections is the TCP/IP architecture. Given the explosion of TCP/IP networking, and the dramatic rise of Internet use (which is based on the TCP/IP architecture), this topic is worth further exploration.

TCP/IP intranetwork gateways

In order to appreciate the role of TCP/IP gateways, we must first explore the TCP/IP network address strategy. Under the TCP/IP architecture, each system in the network is assigned a four-byte (32-bit) address, termed the IP address. The IP address is normally represented as $w.x.y.z$, where w, x, y, and z are replaced with a decimal number between 0 (hex 00) and 255 (hex FF). For example, 192.0.0.12 is a valid IP address.

The four-byte IP address is further broken down so that a portion of it identifies a logical network address and the rest of it identifies a system within that network. For example, 192.0.0.12 identifies a specific system (address 12) within a specific network (address 192.0.0). Similarly, address 128.10.20.12 identifies system 20.12 within network 128.10. The IP address can be composed using one of three different formats, termed *classes*, which break the overall address down into network and host system components differently.

☐ Class A—Uses the format *network.host.host.host*, with the network component falling in the range between 0 and 127 (exclusive of 0 and 127), and the host component being greater than 0. For example, in the address 64.0.1.12, 64 identifies the network, and 0.1.12 identifies the host system.

☐ Class B—Uses the format *network.network.host.host*, with the first network component falling in the range from 128 to 191 (including 128 and 191), and the host component being greater than 0. For example, in the address 130.101.0.68, 130.101 identifies the network, and 0.68 identifies the host system.

☐ Class C—Uses the format *network.network.network.host*, with the first network component falling in the range from 192 to 223 (including 192 and 223), and the host component being greater than 0. For example, in the address 200.1.1.37, 200.1.1 identifies the network and 37 identifies the host system.

IP addresses define boundaries for different physical networks, and in some network implementations these addresses also create barriers between logical networks. In many cases, a system at address 192.0.0.68 cannot communicate with a system at address 193.0.0.6, even if the two systems are physically on the same cable segment. In general, it is safe to assume that every system must go through a gateway to access a system residing in a different physical or logical network.

Gateways communicate with one another to share routing information and communicate changes that occur to the routing information. For example, if a gateway picks up a new network connection, it can share this information with other gateways. In the world of TCP/IP, gateways are divided into three categories: interior gateways, noncore exterior gateways, and core exterior gateways. These are shown in Fig. 5-4, and are defined as follows:

☐ An interior gateway exists inside an autonomous, self-contained network. An autonomous network may be a small local LAN, a series of interconnected LANs, a campus-wide network, or anything in between. The purpose of an interior gateway is to provide interconnections within the confines of the autonomous network.

☐ A noncore exterior gateway, on the other hand, provides a connection between two or more autonomous networks. In other words, an exterior gateway interfaces an autonomous network to the outside world. Exterior gateways can link two autonomous networks or connect an autonomous network to the Internet.

☐ A core exterior gateway provides interconnections within the framework of the Internet network. Unlike noncore exterior gateways, core gateways do not interface with autonomous networks—they are dedicated to maintaining interconnections inside the Internet.

The actual protocol used by a gateway to disseminate and maintain routing information is based on the type of gateway it is.

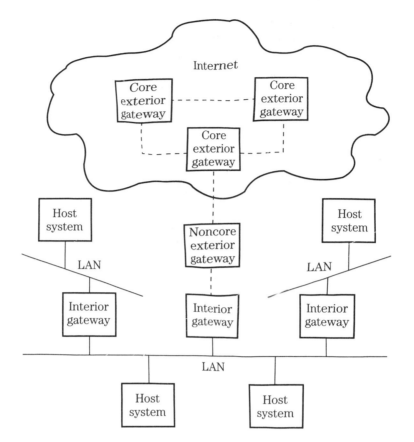

■ **5-4** *Types of TCP/IP gateways*

☐ Core exterior gateways use a gateway protocol known as
   SPREAD to talk with one another. This protocol is only used
   within the Internet. Because core exterior gateways also
   communicate with noncore exterior gateways, they support
   the same protocol used by noncore exterior gateways.

☐ Noncore exterior gateways use a protocol known as the
   Exterior Gateway Protocol (EGP). This protocol can be used
   between any noncore exterior gateways to interconnect
   autonomous networks, or between a noncore gateway and a
   core exterior gateway to connect an autonomous network to
   the Internet. A noncore exterior gateway must also support at
   least one interior gateway protocol, so that it can
   communicate with interior gateways or host systems.

☐ Interior gateways support a variety of protocols. This variety
   is available because there are no formal Internet connectivity
   requirements within autonomous networks, so networking

vendors can offer whatever protocol they happen to endorse. Also note that these protocols are used by the host systems to communicate with gateways. Some of the more common protocols are:

~ HELLO—The HELLO protocol is one of the original interior gateway protocols used in TCP/IP networks. HELLO is different from the other interior gateway protocols in that it determines the best route for a message to take based on the estimated elapsed travel time. Although the HELLO protocol is still in use, it is not nearly so widespread as RIP.

~ RIP—The Routing Information Protocol (RIP) was developed to provide routing services in Xerox Network System (XNS) networks. RIP operates at the TCP level of TCP/IP and assumes a path that passes through the fewest routers (has the fewest hops) is the best path to take.

~ OSPF—The Open Shortest Path First (OSPF) protocol was developed to replace RIP, and offers a number of advantages over RIP. First, it operates at the IP level within TCP/IP, so it is capable of routing a wider variety of services than RIP. Second, OSPF is not as "boisterous" as RIP—under OSPF, routing table updates get distributed only to neighboring routers. Finally, OSPF takes into account a number of factors to determine the best possible route, including line speeds, error rates, and current loading.

Please note that the RIP and OSPF protocols are in fact the same protocols used in routers, as discussed in Chapter 4.

Given that TCP/IP gateways use "routing" protocols, you may be wondering how gateway functions differ from the functions offered by routers. The simple truth is that they do not. Gateways are called "gateways" in TCP/IP networks because they were around long before routers came into popular use. Please refer to the section entitled *Gateways versus routers* for more information on this subject.

Interoperability gateways

Gateways in the "interoperability" category implement application-oriented services to provide interoperability between dissimilar networks. This is the type of device that most people picture when they think of gateways.

Services that are often offered by this class of gateways include

☐ terminal emulation—The ability of a terminal to appear to a foreign host as a native terminal. For example, an IBM 5250 workstation might appear to a Digital VAX system as a Digital VT 220 terminal.

☐ file access, a capability that can take on many forms:
  ~ The simplest implementation of file access is file transfer. In this case, a file is physically copied (and translated between ASCII and EBCDIC if necessary) from one system to another. For example, this is how the TCP/IP File Transfer Protocol (FTP) and the IBM Remote Job Entry (RJE) facility operate.
  ~ Record-level access on a remote file allows a program to write and read a predetermined file as if it were a local file. For example, in the Digital VAX DECnet and IBM AS/400 environments, specific files on remote systems can be opened for read/write access.
  ~ File set access allows one system to mount a group of files on a remote system and access them as local files. File set access includes record-level access. This is the principle behind the TCP/IP-oriented Network File Services (NFS), Novell NetWare, Digital Pathworks, IBM PC Support, IBM CA/400, and other similar products.

☐ electronic message exchange—Support for the interchange of messages between electronic mail packages. For example, IBM OfficeVision products or Digital's All-in-1 package might exchange mail with UNIX Simple Mail Transfer Protocol (SMTP) programs.

☐ network management reporting—The translation of one type of network management input into another. For example, Simple Network Management Protocol (SNMP) data might be translated into SNA NetView format.

☐ program-to-program communications—The facility for programs on dissimilar systems to communicate with one another in real time. For example, an IBM LU 6.2 (APPC) program might communicate with a Digital task-to-task or TCP/IP socket application.

☐ resource sharing—The ability to share printers, tape drives, or other peripheral devices. For example, UNIX print queues might be redirected to IBM or Digital print queues (or vice versa).

Rarely does one gateway offer this complete suite of services—most are focused on a smaller set of core services, as shown in Fig. 5-5.

Gateways that offer interoperability services are important tools for implementing distributed processing and cooperative processing environments. Using this technology, the resources and processing capabilities of different networks and different types of systems can be intertwined to create data processing solutions for

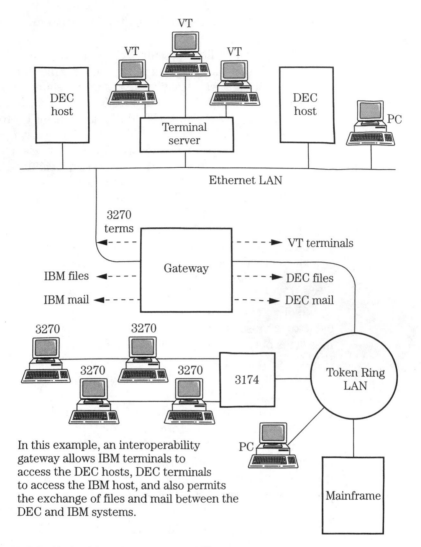

In this example, an interoperability gateway allows IBM terminals to access the DEC hosts, DEC terminals to access the IBM host, and also permits the exchange of files and mail between the DEC and IBM systems.

■ **5-5** *Limited function interoperability gateway*

difficult business problems. For example, technical data can be married to administrative data, expensive printers can be shared between networks, and enterprise-wide mail can span multiple electronic mail packages.

Protocol converters vs. interoperability gateways

By necessity, gateways that fall into the interoperability class perform protocol conversion and device emulation. This puts them dangerously close to the category of devices known as *protocol converters*. Protocol converters have been a means of connecting dissimilar systems throughout the history of data processing. For example, RJE protocol converters have been used to facilitate file transfer with ASCII systems for over 15 years.

As data processing has become more complicated, so have the needs for multivendor connectivity solutions. The low speed file transfer solution of yesterday is often not enough of a solution today. The market has come to demand high speed file transfer, mail exchange, terminal emulation, and more. Gateways have evolved from the genetic pool of protocol converters to address these requirements.

Because of this genealogy, it is often difficult to separate protocol converters from gateways. There are some general guidelines that can help you make the distinction, however:

☐ Gateways generally have a level of intelligence above and beyond protocol converters. Gateways use general-purpose CISC or RISC processors, while protocol converters use custom circuits and device-oriented processors (like the Zilog Z80 or Intel 80186).
☐ Gateways are normally implemented to facilitate multiple functions or services. For example, a single gateway can handle terminal access, file transfer, and program-to-program communication. In contrast, protocol converters are usually fixed-function and handle one task (terminal emulation, for example).
☐ Gateways usually cost more than protocol converters. Using a general-purpose processor to perform multiple functions tends to cost more than a device-oriented processor performing a fixed function. In a gross sense, it's like comparing a PC to a calculator.

As the costs for general-purpose processors have spiraled downward, the ranks of low-end gateways have swelled and the ranks of protocol converters have diminished.

Are gateways hardware or software products?

Many people think of a gateway as a hardware device. The simple truth is that a gateway must be a combination of software and hardware. The question becomes "Whose hardware do you want to use?"

The gateway vendor community is very divided on this subject. Some vendors believe that the best solution is to provide hardware that has been especially selected or optimized for the task. Other vendors believe that the best approach is to load software on an application system that interfaces with a network adapter. And a third set of vendors believe that you should do a little of both—use an independent hardware box in conjunction with application system-based software.

Each philosophy has its pros and cons. But no matter how you analyze it, it all boils down to balancing two issues: cost and resource utilization. Obviously an independent gateway can cost more than a software-based solution, but at the same time, a software-based solution can consume CPU, disk, and memory resources in your application systems.

Sometimes you have no choice about the architecture. For example, if you're looking at tunneling SNA through TCP/IP, you'll find most solutions are hardware-based. In those cases where you can choose an approach, you should take a hard look at both your budget and your application resources. Do you want to purchase a system from a vendor that is ready to go when it is dropped into your environment? Would you rather install software (and possibly additional interface hardware) on one of your own systems? Or do you want to pull together different software and hardware components and create your own gateway? The choice is often yours to make.

Gateways vs. routers

As previously noted, transport and intranetwork gateways are suspiciously similar to routers. Like routers, they work at the protocol level to control how information flows between multiple networks. Unfortunately, no clear rules or standards are in place that govern the use of the terms *router* and *gateway*.

Historically, the use of proprietary specifications instead of open specifications has been the main factor that has separated gateways from routers. Routers tend to use well-defined protocols like RIP, OSPF, and DLSw, so it is often possible to mix and match routers from different manufacturers in the same network. Gateways, on the other hand, often use private, proprietary protocols or techniques to handle functions like encapsulation, tunneling, or internetwork relaying. This means that it is difficult, if not impossible, to interconnect gateways from different manufacturers.

As we move farther into the arena of standards-based computing, the number of proprietary gateway protocols is declining, and the number of standard-based routing protocols is increasing. Some network architectures, such as TCP/IP, continue to use the term *gateway* to describe a network device that is, for all intents and purposes, a router. The bottom line is that you shouldn't be surprised if you see both terms—*gateway* and *router*—applied to products that perform the same function: you need to learn to look beyond the name and look at the functions and protocols supported by these products. Only then can you determine their fitness for use in your network.

Hubs

According to the dictionary, a hub is "the center of activity" or "a focal point." In networking terms, a hub is the nucleus for multiple wide area and/or local area connections—all roads lead to the hub (sooner or later). This does not, however, mean that a hub is simply a wiring concentrator (although that is a function hubs perform). Today's hubs can include a combination of several types of devices, all neatly packaged together. For example, hubs may perform any or all of the following functions:

☐ Wiring junction—One of the most important aspects of a hub is that it represents a single point where network wiring comes together. For serial lines, a hub might include a multiplexor or a patch panel. For LAN connections, a hub acts as a repeater (Ethernet) or MAU (Token Ring) to provide connections to multiple segments or lobes. Hubs are available to handle a wide range of wiring types, including coaxial, fiber, shielded twisted pair (STP), unshielded twisted pair (UTP), and of course the various IBM cable types (type 1, type 2, type 3, and so on).

☐ Bridging—In a LAN environment, hubs cannot only manage multiple segments/lobes in one logical network, but can also interconnect multiple logical networks through bridge technology (as discussed in Chapter 4). This can be applied in two different ways:

~ Local bridging allows for the interconnection of multiple local networks. For example, two Ethernet networks or two Token Ring networks can be interconnected at the same location. More sophisticated bridging techniques also allow LANs of different types (FDDI, Ethernet, Token Ring, etc.) to be interconnected.

~ Remote bridging allows for the interconnection of remote networks with local networks. For example, a Token Ring network in Walla Walla, Washington can be bridged to a local ring. Remote bridging also involves connection to the WAN environment. Thus, a hub adapter that performs this function must also deal with different types of wide area connections

117

(analog, digital, and packet switching, to name a few). In some cases, hubs may also be able to bridge non-LAN traffic into the local LAN environment (as in the case of bridging SDLC into a Token Ring environment).

☐ Routing—As discussed in Chapter 4, routing is similar to bridging, but routing selectively controls which traffic moves from one network to another. Routing is normally a protocol-specific function. For example, the Novell IPX traffic from a local Token Ring may be routed to a Token Ring network in New York, while the TCP/IP traffic on the same local ring is routed to a ring in Los Angeles. As in the case of bridging, routing may be performed between local or remote networks (but remote is the norm).

Figure 5-6 shows both ends of the hub spectrum. As the hub market grows, more and more functions find their way into hubs. Modem, DSU, protocol converter, and gateway technologies are all on a path leading to incorporation into hubs. With this in mind, the list of hub functions will continue to expand to meet the needs of the networking market.

Hub advantages

In most implementations, a hub is a single unit that includes a backplane (or bus) into which you fit adapters. Each adapter (or set of adapters) performs a specific function (e.g., wiring concentrator, bridge, or router). The shared bus/backplane allows data to quickly and easily move from one adapter to another inside the hub.

For example, in a Token Ring environment, you can have one card to bridge Token Ring over a digital link, another card to bridge multiple local rings together, and yet another card to interface an SDLC line directly into the ring environment. When traffic needs to flow from one adapter to another, it simply travels over the internal bus, thereby eliminating the need for external cables to interconnect related adapters.

The advantages of using hub technology, as opposed to employing a series of independent networking devices, include

☐ Common chassis/common power—By using a single hub with common multiple functions, you have the advantage of using one cabinet (or chassis) and one power source. On a nontechnical front, this also avoids the occasionally ugly result of placing devices of different color and cabinet styles next to one another.

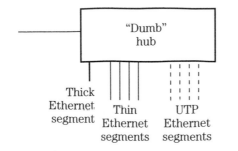

In simple terms, dumb hubs are little more than wiring concentrators. Smart hubs, on the other hand, combine multiple networking functions.

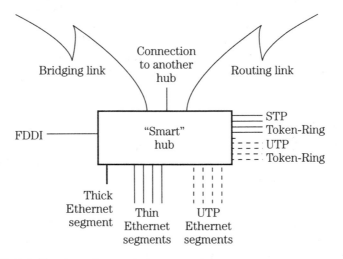

■ **5-6** *Dumb and smart hubs*

☐ reduced cabling—When you employ multiple devices in a single network, you must cable the devices together. This often forces you to scramble to make awkward or mixed-media connections. With a hub, however, the backplane/bus provides the connection between related functions.

☐ single source purchase/support—Because the hub manufacturer provides all of the functions, you have one source for purchasing and one source for support. In more complicated networks, this reduces opportunities for finger-pointing among networking device vendors.

☐ consistent operational interface—Setup, operation, and maintenance of the hub will be consistent, regardless of what types of adapters are added or removed. With multiple networking devices, each device potentially has a different and unique operational interface.

☐ integrated network management—Most hub vendors support either the Simple Network Management Protocol (SNMP) or the Common Management Information Protocol (CMIP). This allows one protocol to be applied to all of the functions within the hub. If you employ independent devices, you may be faced with a series of devices that use different (or no) network management protocols.

Also note that hubs are not necessarily tied to any single LAN or WAN environment. For example, a single hub can often handle Ethernet and Token Ring LANs, and can optionally bridge or route traffic between the two. Hub adapters are also available to handle the burgeoning numbers of metropolitan and wide area connections, such as Fiber Distributed Data Interface (FDDI), Asynchronous Transfer Mode (ATM), Switched Multimegabit Data Service (SMDS), frame relay, and others. These technologies are discussed in the second half of this book.

Hub disadvantages

Hub technology is extremely flexible, and adapters are available to perform a wide variety of functions. But no technology is perfect. The downside of employing hubs includes

☐ price—Although hubs are very cost-effective on the low end of the scale, entry-level pricing for high-end hubs is high—mostly because you're paying for the capabilities of the chassis, even though you won't fully use it initially. Once you fill a hub with a diverse array of adapters, however, it's normally cost competitive with using independent devices.

☐ single point of failure—If a hub goes down, your network is in trouble. This is especially true if your hub is crammed full of concentrators, bridges, and routers. Hub vendors are sensitive to the issue of downtime, and take extra measures to ensure continued operation, including, for example, redundant power supplies. Several vendors also allow you to add, remove, or swap adapters while the hub is operating.

☐ limited to capabilities of the hub vendor—Hub adapters are not generally compatible from one vendor's hub to the next. This means that the capabilities of your hub are limited by the offerings of the hub vendor. If, for example, the hub vendor you selected does not make an Ethernet/Token Ring bridge adapter and you desperately need one, you will be forced to bring in an external device to handle that function. Once you open the door and let in additional standalone networking devices, you start detracting from some of the advantages of a hub (single source, consistent interface, common network management protocol, etc.).

Most of the negatives associated with hubs can be overcome or avoided by careful shopping. If you understand your current network environment, and can anticipate both your short-term and long-term network needs, you stand an excellent chance of choosing a hub vendor who can stay with you for the long run—or at least until the next acquisition or change in business plan.

121

IV

High-speed
interconnections

ONCE ETHERNET AND TOKEN RING LANS BEGAN PROLIF-
erating, people began to extend and interconnect those LANs. At
the start, bridges and routers satisfied the need to bring depart-
mental LANs together, while telephone services provided wide area
interconnects. However, these traditional wide area networking
techniques have intrinsic problems. First, although bridges and
routers operate at the same speed as their client networks, they
tend to become bottlenecks, due to interactions of the combined
network traffic loads. Telephone interconnects (generally limited
to 56 kbps) exacerbate the bottleneck problem. Second, interconnect failures are often hard to diagnose, and once diagnosed, hard
to repair. Replacing and reconfiguring a router, or digging up a
buried cable, is a lot more work than rebooting a beaconing workstation. Frequent and long interconnect outages deny users access
to network resources, costing an organization dearly in time and
money. Third, network administrators keep pushing out the physical distances between LANs, from interfloor, to interbuilding, to in-
tercampus, and eventually intercity and intercountry. Fourth,
growing Ethernet or Token Ring LANs eventually "max out," hitting
either node or traffic limits that severely degrade performance.
Fifth, and finally, new applications like video conferencing require
real-time data transfer that can't abide propagation delays that
occur in traditional LAN interconnects. These five limitations—
bandwidth, availability, distance, performance. and propagation
delay—spurred the development of high-speed interconnect tech-

nologies. Out of that development, two kinds of standards emerged: FDDI (Fiber Distributed Data Interface) and fastpacket.

FDDI (pronounced "fiddey") is the natural successor to Ethernet and Token Ring. It's a LAN, rather than a WAN, technology; you can use it as a backbone between lower-speed LANs, or as a single large LAN by itself. FDDI uses optical fiber to carry data at 100 Mbps between nodes as far apart as 60 km, supports up to 500 nodes, and permits network route distances totaling 100 km. These properties address the bandwidth, distance, and performance problems. Optical fiber itself is immune to electrical and RF interference. This property makes FDDI media more reliable than electrical media, and combined with FDDI's provisions for self-healing networks, addresses the availability problem. Finally, FDDI's architecture supports real-time traffic through its synchronous service, which guarantees priority traffic delivery even on a heavily loaded network.

FDDI uses dedicated fiber cables, either installed by you or leased from a communications company. This makes FDDI an easy buy for geographically small networks, but a major undertaking for networks that must traverse others' property. Leased fiber (called "dark fiber" because you supply the light signal) is expensive, and rolling out your own cable is even more expensive. If you're not General Motors or IBM, you will probably need an alternative to FDDI technology for long distance and off-campus interconnects.

That alternative is the packet switching technology called fastpacket. For network builders (as opposed to telephone companies), this usually means fastpacket services, since only the largest companies are likely to actually build their own fastpacket networks. Asynchronous Transfer Mode (ATM) is the umbrella standard for fastpacket technologies, of which there are currently three: Frame Relay Service (FRS), Switched Multimegabit Data Service (SMDS), and Cell Relay Service (CRS). You buy these services in bandwidth increments from a communications vendor. For dedicated interconnects appropriate to linking LANs, CRS offers the most bandwidth bang for the buck, and is discussed here in Part IV on high-speed interconnections. Point-to-point and point-to-multipoint interconnects appropriate to non-LAN applications such as electronic data interchange (EDI) and digital video broadcast are discussed in Part V under multipoint links.

ATM addresses the aforementioned bandwidth and distance problems handily by supporting two speeds, 155 Mbps and 622 Mbps, and virtually unlimited distances—distance concerns are the worry of the service supplier. The supplier, generally a Local Exchange Carrier (LEC) or other large communications service vendor, has the ability to switch to backup routes and equipment instantaneously, addressing the availability problem. ATM jumps performance and propagation delay limitations by providing for bandwidth-on-demand and prioritized traffic: when network traffic requires additional bandwidth, or a particular service (teleconferencing, for example) needs real-time delivery, ATM allocates the necessary resources to meet those needs. The service provider may bill a surcharge in such situations, but that's usually a lot cheaper than deploying a private physical WAN capable of handling your peak traffic requirements.

125

Fiber links & FDDI

COPPER WIRE AS A LAN COMMUNICATIONS MEDIUM HAS ITS problems:

☐ Distances are limited to building-sized routes.
☐ Physical laws put a practical ceiling on bandwidth.
☐ Long distances cause network timing problems.
☐ Radio, electrical, and crosstalk interference garble data.
☐ Copper's conductivity propagates lightning and surges.
☐ Harsh environments reduce reliability.
☐ Electromagnetic emissions compromise security.

As networks grow and interconnect, these problems loom large. You're likely initially to encounter the first two problems while trying to expand your burgeoning network. You'll push segment lengths to the max, to add "just one more" node or station. This invariably brings about the next two problems, as longer segments have longer signal propagation times and are more susceptible to RFI. Down the road, as your network ages, the third pair of problems will strike. The resulting network failures will be intermittent, puzzling, and difficult to diagnose. And even if you're never unfortunate enough to fall victim to the last problem, just the thought that someone can suck valuable data out of your network without even touching a wire will keep you awake at night.

An alternative to copper, optical fiber, carries data on light waves over a glass fiber rather than as an electrical signal over wire. Optical fiber has advantages that counteract copper's problems:

☐ Optical fiber's low signal attenuation permits distances—up to 60 km—that let you extend networks to campuses of arbitrary size, and even to nearby cities.
☐ Fiber offers virtually unlimited bandwidth potential. The limiting factor in fiber bandwidth is the quality and sensitivity of the sending and receiving units; fibers themselves can be made today to carry ten terabits per second—enough to

transmit all the digital data communicated throughout human history in a single hour (plus or minus a few minutes).

☐ Light waves are immune to radio and electrical interference, and do not interfere with each other across cables. This completely eliminates RFI and crosstalk problems in network wiring, eliminating the special installation and routing considerations which enslave copper.

☐ Glass fiber is nonconductive, so fiber segments don't act as conduits for lightning strikes or high voltages escaping from damaged equipment. This property isolates catastrophic events, rather than propagating them, improving overall network survivability.

☐ Glass fiber is chemically inert, as is its sheathing. Water, heat, and corrosive atmospheres that eventually destroy copper cable have little or no effect on fiber.

☐ The light signal in a fiber cable is completely self-contained, unlike the electromagnetic signals in copper cable, which can be detected and recorded at a distance. The only way to tap a fiber cable is to insert a splice—a nontrivial task.

Beyond these compensating advantages, fiber has unique properties that make it easier and less expensive then copper in many applications:

☐ Glass fibers are made from silica sand, a substance much more abundant on the earth than copper. Although fiber costs more than copper cable today, the price gap is steadily narrowing. Eventually fiber will be cheaper than copper cable.

☐ Fiber cable weighs much less than copper cables having the same data carrying capacity. This simplifies fiber installation and even makes possible new applications in weight-sensitive environments such as aircraft and automobiles.

☐ Fiber cable is more flexible than either coaxial cable or shielded twisted pair (STP) cable. Fiber cables also are much thinner than copper, and thus easier to install. The total data carrying capacity of existing wiring tunnels is much higher with fiber than copper.

☐ Fiber requires much lower power levels than copper to propagate equivalent data signals. While the cost of power itself isn't a significant issue in networking, power expended by network devices—adapters, repeaters, routers and bridges—is the main factor determining the size and power supply requirements of these devices.

Fiber optic networking itself is new and unique enough to warrant an introductory discussion of the physical processes and technology involved.

Fundamentals of fiber optic links

All fiber-based data communications use the same physical processes and technology. Understanding the fundamentals of that technology is important in planning fiber installation and selecting proper components. Those fundamentals include the parts of an optical link, how light propagates in a fiber, and the differences among fiber types.

Parts of an optical link

Figure 6-1 illustrates the parts of a fiber optic communication link. An electrical data signal enters a driver, which controls (modulates) a source of light to produce a modulated light signal. The light signal travels over the fiber until it reaches a light sensor at the far end, called a detector. The detector converts (demodulates) the light signal into a very small electrical signal, which an amplifier then makes into a usable electrical signal. The driver—source parts are collectively called the transmitter, and the detector—amplifier parts collectively make up the receiver.

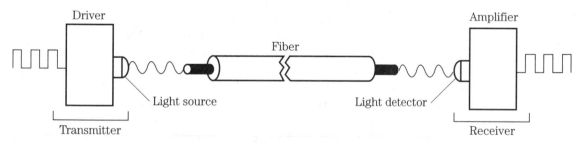

■ **6-1** *Parts of a fiber optic communication link*

In fiber links for data, the light source is usually a solid state device, such as a light-emitting diode (LED) or semiconductor laser. The detector is a photodiode, and the driver and amplifier are simple electronic circuits. The fiber running between the transmitter and receiver is the one mysterious component, and also the one miracle that makes optical communication at all practical. A light beam entering one end of the fiber travels to the other end virtually unaffected by tortuous twists and turns the fiber takes en route.

An optical fiber is a thin, continuous, glass (or sometimes plastic) filament. Its thinness (thinner than a human hair) is what makes it flexible. The fiber itself is really two fibers in one: an outer cladding and an inner core. The fiber is encased in a plastic sheathing for protection and to limit the radius through which the overall cable is easily bent; sharp bends will break the fiber thread, or severely attenuate the signal, as described in the next section. Figure 6-2 is a magnified view of the end of a fiber.

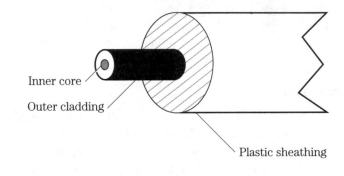

Inner core

Outer cladding

Plastic sheathing

■ **6-2** *Magnified view of a fiber cable*

Light propagation

Light propagates through a glass fiber using total internal reflection, which depends on the fact that the speed of light depends on the medium through which it travels. The outer cladding and inner core have different refractive indices, and the boundary between the two materials acts like a mirror when light strikes it at a shallow angle of incidence (the refractive index of a medium is the ratio of the speed of light in a vacuum to the speed of light in that medium). Light rays enter the core of the fiber at all angles, some steep, some shallow. A steep ray passes through the core boundary and is lost in the cladding; a shallow ray bounces off the core boundary to the opposite side of the core, where it strikes the boundary again at an equally shallow angle. In this way, the ray propagates along the fiber. As long as the fiber contains no sharp bends, the ray continues to the far end of the fiber, where it exits.

When light travels through any medium other than a vacuum, a gradual energy loss—called attenuation—occurs as the medium absorbs some photons. In optical fiber, attenuation is measured in decibels (dB). When planning fiber networks, you compute the total attenuation over a fiber in dB to ensure that the light at the far end of the fiber retains enough energy to be amplified by the receiver. Fiber cable suppliers provide tables for making such computations, taking into account the particular characteristics of a given type of cable.

Types of fiber

In data communications, fibers come in two flavors: single-mode and multimode. The word mode in this context is synonymous with ray; you can think of a single-mode fiber as transmitting a single ray, and a multimode fiber as transmitting multiple rays. Figure 6-3 illustrates the differences between the two types. The physical difference is that single-mode has a small core in relation to the diameter of the cladding, while multimode has the reverse. The important functional difference between the two types of fibers is simple: single-mode fibers require a stronger light source and have lower attenuation, and can thus carry the light farther; multimode fibers use a weaker light source and carry a signal a shorter distance. Single-mode fiber works without repeaters for distances of up to 60 km. The two types of fibers cost about the same, but transmitters and receivers for single-mode fiber are considerably more expensive than their multimode counterparts, due to their high-powered light sources and tighter mechanical tolerances. When selecting cable, you'll choose single-mode fiber for long hauls (>2 km) and multimode fiber for short hauls.

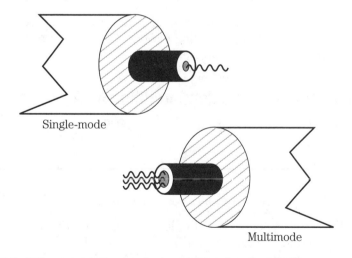

Single-mode

Multimode

■ **6-3** *Differences between single-mode and multimode fiber*

Types of transmitters & receivers

The light source in a transmitter can be either an inexpensive LED or a much more expensive semiconductor laser, also called a laser diode. Laser transmitters, with power output measured in milliwatts, are only necessary for long-haul segments, usually over single-mode fiber. This is a good thing, as such transmitters are often two to five times as expensive as LED-based transmitters. LEDs emit less than a hundred microwatts, which greatly reduces the complexity of the driver circuitry, the amount of heat generated, and the failure rate. Laser sources also can present a safety problem, as the lasers used for full-distance single-mode fiber links can cause eye damage. Such equipment requires special safety precautions that must not be taken lightly. For this reason, you'll find power-limited transmitters on the market that don't support the full distance capabilities of single-mode fiber but that also don't require special handling.

The light detector in a receiver is one of two types of photodiode: PIN or APD. The PIN detector's name describes its construction: p- and n-type doped semiconductors sandwiching an intrinsic semiconductor. The APD (avalanche photodiode) detector is so-called because it uses a phenomenon called the avalanche effect to detect extremely small variations in light intensity. APDs are more sensitive than PINs, and of course more expensive. APDs also require a 100 Vdc power source, which makes for bulkier and more expensive receiver circuitry.

The components and principles described above give you a foundation for understanding various applications of fiber links. Now let's look at those applications.

Using fiber links to extend traditional LAN links

There are three routes to gaining fiber's advantages in existing networks:

- ☐ Extend traditional LAN segments using fiber optic links.
- ☐ Install a fiber optic backbone between LANs.
- ☐ Replace part or all of an existing LAN complex with fiber.

The first route uses simple point-to-point adapters and cable; the second and third routes require FDDI (Fiber Distributed Data Interface) technology, discussed in the next section.

The most common use of point-to-point fiber is to extend an existing network segment when copper's distance abilities prove inadequate. Such links can extend distances between hubs or segments in Token Ring and Ethernet networks, as long as the extension is between networks of the same type. You could, for example, extend a Token Ring network in one building to a Token Ring network in another building. Such an extension is called a Fiber Optic Inter Repeater Link (FOIRL). Figure 6-4 shows the components required for a FOIRL. To transition between the existing LAN cabling and the fiber cable, two transceivers are necessary, one at each end of the fiber extension. A transceiver is a self-contained unit comprising a transmitter and receiver and the necessary supporting circuitry, and requires a separate power supply (so think twice before you locate your fiber transceivers out in a parking lot somewhere).

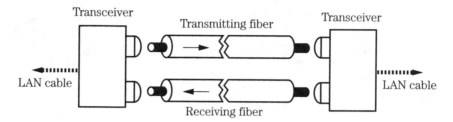

■ **6-4** *Fiber Optic Inter Repeater Link (FOIRL) components*

One side of a transceiver provides an electrical connection to the traditional network. The other side has connectors for two fiber cables, one to send, the other to receive. Unlike copper, which is a broadcast medium carrying signals in both directions, optical fiber carries a signal in one direction only.

The transceiver itself is a model of simplicity: you connect it to the LAN, the fiber, and power, and it works. There usually are no switches to set or parameters to configure. The only weak link in this scenario is the human being setting up the link, who can make the rather subtle mistake of misconnecting the fiber cables. Many transceivers use separate cables and connectors for the send and receive fiber. Figure 6-5 illustrates a few of the distressingly unstandardized possibilities. Connecting fiber cables backward on one end can damage the light source components, particularly in laser-equipped transmitters. The only guaranteed preventive measures are diligence and the use of color-coded cable and connectors.

A significant problem with FOIRL is the lack of standards in cabling, connectors, and transceiver compatibility. Transceivers from vendors often don't interoperate, and often have incompatible connectors and fiber requirements. This usually means buying point-to-point products in matched pairs from the same vendor. This is tolerable when the only goal is extending an existing LAN

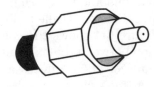

SMA 906

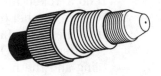

Biconic

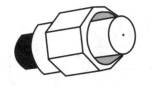

SMA 905

ST

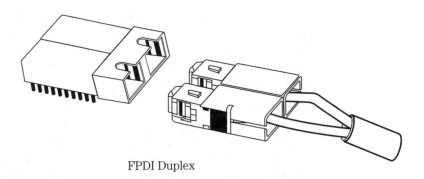

FPDI Duplex

■ **6-5** *Variety of fiber optic cable connectors*

segment. Using fiber to interconnect many LAN segments, or multiple LANs, requires better standardization.

Another problem with FOIRL is signal delays that interfere with the extended network's access method. For example, an Ethernet network requires that collisions be detected while a node is still transmitting a packet. Long fiber extensions can delay the collision notification so long that the transmitter never sees it, or worse, sees it when transmitting the next packet. This can result in excessive retries causing slowdowns or connection timeouts. Token Ring has similar problems with token rotation time (TRT): a long fiber extension can increase TRT to the point where network performance degrades unacceptably. For Ethernet, the sum of all fiber spans should not exceed 2500 meters. Token Ring can go considerably farther, the total distance depending on the number of active stations in the ring. FOIRL component vendors provide formulas for sizing both Token Ring and Ethernet extensions.

The Fiber Distributed Data Interface

The standardization missing in FOIRL is found in FDDI, the Fiber Distributed Data Interface. As discussed in the introduction to Part IV, FDDI is a full-fledged LAN standard, running at 100 Mbps, and based on Token Ring mechanics. It breaks traditional LAN topology barriers, supporting networks of up to 500 nodes with segment lengths as long as 60 km. The total route distance can reach 100 km. Besides high bandwidth and large area coverage, FDDI has several important advantages over traditional LANs designed into it:

☐ Fault tolerance. An FDDI network can be designed with a redundant secondary path that automatically heals the ring in the event of single-segment failures, without losing any active stations.
☐ Support for real-time traffic. Unlike Ethernet and Token Ring, where congestion delays make real-time applications such as voice and video impractical, FDDI's synchronous service guarantees on-time delivery for high-priority traffic.
☐ Voice/data/video integration. Along with guaranteed delivery time, synchronous service offers the ability to partition bandwidth so that real-time applications have a guaranteed throughput when necessary.
☐ Distributed network management. All network monitoring and control algorithms operate in a distributed way among the active stations. There is no need to elect an active monitor, as with Token Ring.

These features arrive on the LAN scene just as users need additional bandwidth for new applications. To make FDDI easier to mix with existing networks, a companion standard called CDDI (Copper Distributed Data Interface) supports the full set of FDDI protocols over 100-Mbps copper cable. CDDI is described in more detail later.

135

History of FDDI

FDDI is actually a group of standards, produced by the American National Standards Institute (ANSI) X3T9.5 task group over a ten-year period. Figure 6-6 lists the important FDDI historical milestones.

October	1982	First consideration in ASC X3T9.5
December	1982	FDDI name chosen
June	1983	Detailed proposal submitted
October	1984	X3T9 technical letter ballot for Media Access Control (MAC)
July	1987	First ANSI standard for FDDI MAC
February	1989	X3T9 technical letter ballot for FDDI-II
April	1989	First ISO standard for FDDI physical layer
July	1990	X3T0 technical letter ballot for Station Management (SMT)
June	1992	SMT forwarded to X3 for approval

■ **6-6** *Important FDDI historical milestones*

The standards cover physical hardware, including cables and connectors, and the physical and data link protocol layers. They also include a Conformance Testing standard against which vendors can measure their product to help ensure compatibility.

The current standard, also known as FDDI-I, provides for basic data-only operation. A second standard, FDDI-II, is in the works to support hybrid data and real-time applications through the provision of isochronous service. Isochronous means literally equal time; isochronous services guarantee certain applications access that appears to be dedicated as far as timely packet delivery is concerned.

These two standards should keep users reasonably happy until the next LAN standard comes along. That standard is in the works, as a project called FDDI Follow-On LAN (FFOL). FFOL is a one-gigabit-per-second network, and like FDDI, will initially be deployed as a backbone as FDDI itself becomes the primary LAN carrier. An alternative LAN/MAN/WAN technology that also uses fiber, Asynchronous Transfer Mode (ATM), is described in detail in Chapter 7.

To understand how FDDI achieves its superiority over traditional LANs, you need to understand the FDDI access discipline, or Media Access Control (MAC).

Access discipline

The FDDI MAC is a token-passing discipline, with many similarities to traditional IBM and IEEE 802.5 Token Ring. (See Chapter 3 for a description of Token Ring principles and terminology.) There is one large difference, though: FDDI uses a timed-token access method.

136

In a traditional Token Ring, the token circulates around the ring (regardless of how the network is physically cabled) and a station wanting to transmit simply holds the token for a fixed time. During this token holding time (THT), the station sends a frame. When the receiver obtains the data, it sets a flag in the frame acknowledging receipt and releases the frame back into the ring. The sender sees that the frame has made it (or not) and removes the frame from the ring, a process called stripping (described in detail in Chapter 3). The sender then generates a new token to allow another system to have access to the ring. The sender must wait to regain the token before sending again; the time required for the token to go once around the ring is called ring latency.

This scheme works well as long as the ring latency is small (i.e., there are few active stations). But with many active stations, the token can take a long time to make the circuit, which keeps the sending station idle. If the sending station has a lot of data to send, much potential ring bandwidth is wasted.

Figure 6-7 shows how the timed-token access method works. At network initialization, the stations negotiate a maximum latency

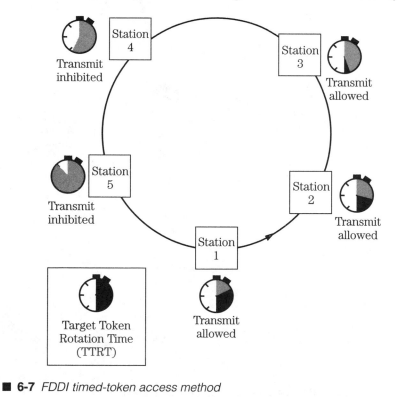

■ **6-7** *FDDI timed-token access method*

time, called the target token rotation time (TTRT). Each station measures the time between arrivals of the token, and if that time exceeds the TTRT, the station doesn't transmit but immediately gives up the token. If the TTRT hasn't been exceeded, the station can transmit for a time up to the remaining TTRT. This has the effect of ensuring that nontime-critical applications, such as file transfer, get to use a fair amount of the ring's bandwidth in the presence of "chatty" applications that only need the network for small or infrequent packets.

The timed-token access method has another performance-enhancing twist, called immediate token release. Figure 6-8 illustrates the

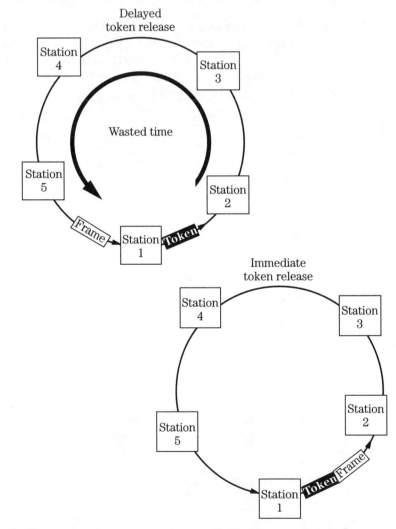

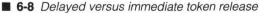

■ **6-8** *Delayed versus immediate token release*

difference between delayed and immediate token release. A traditional Token Ring station holds the token until it has stripped the data frame, wasting the time required for the frame to return. FDDI releases the token immediately after the last byte of the frame, so that other stations can access the token more quickly. Immediate token release is called early token release in 802.5 networks, and is an optional (but infrequently implemented) feature of that standard.

FDDI also supports throughput- and delay-sensitive synchronous traffic, as mentioned earlier. The TTRT accommodates this by guaranteeing that the token will always be available to high-priority synchronous applications after no more than twice the TTRT. You might wonder why, if the TTRT is supposed to be the target rotation time, the best guarantee FDDI makes is double that number. The reason is the so-called yellow light rule, where a station transmitting just before the TTRT expires is allowed to finish sending an entire frame, which results in an overrun of up to the maximum frame size. The algorithm for computing a TTRT takes frame size into account so that the double-TTRT guarantee can be met.

FDDI has a special feature to accommodate bursty asynchronous traffic; that is, intermittent high-volume streams that require several token rotations for complete transmission. A station having bursty traffic can mark the token as restricted. The restricted token can't be claimed by any other asynchronous station, and the token remains restricted until the originator unrestricts it. This mechanism turns out to be a problem, because the token itself doesn't identify its originator. If the originator dies, special recovery mechanisms will eventually pull the restricted token off the network, but repeated failures can seriously degrade throughput. All FDDI implementations currently honor restricted tokens, but few application interfaces support them.

As with 802.5, FDDI supports eight priority levels, 0 through 7, with 0 having the least precedence. However, unlike the strict priority control of Token Ring, FDDI uses a load-level-influenced priority scheme. At any given instant, a station can determine the load of the network as the percentage of available TTRT. (The actual algorithm is more complex, to take into account synchronous traffic, but this is a reasonable approximation.) As loading increases, lower-priority messages are disallowed. At loads above 100%, only synchronous traffic is allowed. Because the actual TRT can be up to twice the TTRT, the load value can be greater than 100%, although for traffic management purposes, only loads up to 100% are meaningful.

At initialization, a given network can configure the number of available priorities (up to eight), TRT thresholds for each priority, and a synchronous traffic TRT allocation. For example, a three-priority system would use priorities 5, 6, and 7, partitioning available TTRT according to load as shown in Fig. 6-9 (note unequal TRT thresholds).

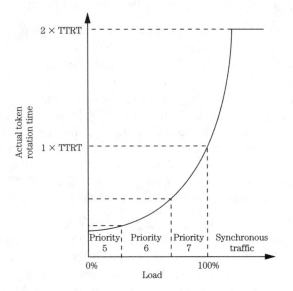

■ **6-9** *Token rotation time as a function of load*

FDDI's token is nearly identical to 802.5's—a three-byte frame:

Start Frame Delimiter—A one-byte pattern indicating the start of the frame.

Access Control—A one-byte field used for control and maintenance functions in normal data messages. The first bit of this field is the "token" bit: when set to one, the frame is a token. The second bit marks the frame as a restricted token. Other bits indicate MAC and station management (SMT) frames.

End Delimiter—A one-byte pattern signaling the end of the frame.

Ring monitoring

Unlike traditional Token Ring, FDDI does not use an active ring monitor. Instead, each station on the ring monitors ring conditions for proper operation. These condition checks are similar to those of a Token Ring elected monitor:

☐ Watch for data frames that travel around the ring more than once, removing them as they are discovered. Frame fragments, error frames, and runaway restricted tokens must also be detected.

☐ Notice the loss of a token, at which point the claim token process is started to recreate the token.

☐ Maintain a sanity timer requiring that a valid frame or token be seen by a certain interval, called the valid transmission timer (TVX) (typically 2.5 ms).

FDDI is lacking one monitor function found in Token Ring: a master clock source slaved to the active monitor to synchronize all stations on the ring. Instead, FDDI permits slight variations in data stream clocking, using an elasticity buffer to accommodate timing differences over the time of a single frame.

Token loss, perpetually circulating frames, and TVX expirations are soft errors that FDDI's cooperative monitoring quickly detects and resolves, usually by initiating the claim token process. A broken ring is considered a hard error that requires a more sophisticated problem resolution process, called beaconing, which is described in detail in Chapter 3.

Station management

Where traditional Token Ring has a set of basic procedures to initialize and test the network, FDDI has a rich suite of protocols collectively called Station Management (SMT). These protocols provide for network initialization and control, performance and reliability monitoring, and fault isolation and recovery. The protocols can be used by any station in the network, thus distributing functions traditionally delegated to a Token Ring active monitor.

SMT functions include

☐ ring initialization, where stations negotiate the ring's operational parameters (such as TTRT) and verify proper operation of the network.

☐ connection initialization, during station attachment, in which a station tests its links to neighboring nodes and exchanges information about itself with its neighbors.

☐ topology control, where various rules about valid station connections are exercised whenever a station connects to the ring.

- status monitoring, in which a map of the ring is constructed for purposes of collecting and exchanging information about performance and problems observed on the ring.
- error monitoring, where bit error rates are monitored to provide early detection of failing stations so that they may be removed from the ring.
- parameter setting, which lets a remote network manager set various station configuration values and possibly change negotiated network values.
- fault isolation and recovery, which automatically locates and reports physical faults, and reconfigures the network around such faults to ensure continuous operation.

FDDI SMT is a complex subject in its own right, and primarily of interest to those constructing network management tools. Most SMT functions run automatically, periodically performing link confidence tests to ensure reliable operation, and taking sensible actions in the face of network faults. The important issue for network implementers is the set of topological rules for FDDI, described in the Topology and Construction section below. If you follow FDDI's rules, you'll reap the benefits of SMT's rich feature set automatically.

Addressing scheme

FDDI follows the IEEE addressing conventions, which are also followed by IEEE 802.5 Token Ring. The FDDI frame format also matches 802.5 (see Fig. 6-10):

- Control flags to indicate if the frame is a token, a network protocol (data) message, or an SMT control message. If the control flag indicates the message is a token, the remaining information is omitted.
- A source address that identifies the system originating the frame.
- A destination address that identifies the system (or group of systems) the frame is intended for.
- An optional source-routing field for backward compatibility with source-routing networks.
- An indication of what network protocol (e.g., TCP/IP, IPX, SNA, DECnet) is carried in the frame.
- A data area that contains the network protocol information along with any "real" data. This network protocol information can also contain high-level addressing information that defines a logical address and logical network for the transmitting and receiving systems.

□ A checksum that enables the receiver(s) to verify that the frame has arrived intact.
□ A status flag to indicate if the frame was received by its intended recipient(s) or not.

Start frame delimiter	Access/ frame control	Destination address	Source address	Source route (optional)	Information field	Frame check sequence	End delimiter	Frame status

■ **6-10** *FDDI frame format*

The source and destination addresses correspond to low-level (MAC-level) hardware addresses of the network interface card, or addresses overlaid on top of the hardware addresses. As in the case of 802.5, these addresses are the "bottom line" for getting a frame to a system.

FDDI uses a 48-bit address similar to 802.5's scheme (Fig. 6-11), although the standard supports 16-bit addresses, which are now defunct. The address is represented as six pairs of hexadecimal digits. Each system must have a unique address to participate in an FDDI network. That address is normally assigned by the hardware manufacturer and programmed directly into the network adapter hardware, but in most FDDI networks, the address is overlaid with a "soft" address, termed the *locally administered address*. These locally administered addresses can have any structure or format except for the first two bits, which are the individual/group (I/G) and universal/local (U/L) flags, respectively.

Individual/ group (I/G) bit	Universal/ local (U/L) bit	Organizationally unique identifier (OUI) 24 bits	Manufacturer-assigned identifier 24 bits

■ **6-11** *FDDI address format*

Addresses belonging to an individual station have an I/G of 0; addresses representing a group of stations (for multicasting) have an I/G of 1. The U/L bit is 0 for addresses that are unique in the known universe (generally understood to be earth and its surrounds). Such addresses are assigned by the IEEE. A U/L of 1 means the address is locally administered, and is unique only within the local environment. As with Token Ring and Ethernet, universal addresses are assigned in ranges by manufacturer. These manufacturer assignments, or *Organizationally Unique Identifiers* (OUI), are too numerous to list; new assignments occur so rapidly that any printed list is soon out of date.

As with Token Ring, the bits in the address are "backwards" with respect to Ethernet ordering. This usually is not a problem in operation, as hardware implementers tend to get the ordering right (if they don't, their products won't work!). For humans, though, the bit-order issue presents a conundrum: when viewing an FDDI address in hex representation, there is no way to know if the system you're using to display the data has correctly transformed the binary address string to a canonical hex representation. Worse, addresses in embedded protocols, such as Ethernet or Token Ring, may have their addresses turned around too. Or they may be turned around twice, which makes them seem correct. It's all very confusing, and the only known solution is to take care when interpreting hex address representations.

FDDI LANs also make use of functional addresses similar to IEEE 802.5's functional addresses. These addresses are used for FDDI operational features, such as SMT functions. Functional addresses set bits 0 and 1 to "1" (resulting in an FDDI canonical hex value of 03). For example, address 03-00-00-00-02-00 is the functional address for SMT network entities.

Routing

When connecting an FDDI network to other networks, either FDDI or non-FDDI, the question of routing comes up. If a frame leaves an FDDI station destined for a station on a Token Ring or Ethernet LAN, some procedure must exist for moving the frame across the intervening network boundaries. The routing problem is this: locate the destination system and determine a reasonable path, across bridges, to that system; then forward frames to the first waypoint (hop) on the route. The two traditional networking camps—Ethernet and Token Ring—have developed different, and incompatible, routing techniques.

Ethernet uses transparent routing, where all bridges on an extended LAN communicate with each other to build a spanning tree, which establishes unique routes between any two stations. The advantage to this technique is that originating stations don't need to know anything about destinations except their address—the routing is done "on the fly" by the routing components in the network. Chapter 2 discusses Ethernet's transparent routing in detail.

144

Token Ring uses source routing, in which the originating station must first find a route to the destination by interrogating the network. The returned route information is then prepended to future frames so that the frame is moved from bridge to bridge, as required, until it reaches its destination The advantage of source routing is that it simplifies bridge logic: source routing bridges do not need to monitor traffic, maintain internal routing tables, or communicate with each other the way transparent routing bridges do. Chapter 3 discusses Token Ring's source routing in detail.

To marry the Ethernet and Token Ring camps, the FDDI standard supports both transparent and source routing. It also supports a new type of source routing technique called source routing transparent (SRT). SRT-compatible bridges use transparent routing by default, but support source routing as an option.

Thus, a station on a Token Ring network can use source routing to communicate with another source-routing station on a far-flung network—FDDI bridges will handle all the routing procedures. Yet an Ethernet station can blast frames into the ether without worrying about routing, and the FDDI bridges will route the frames on the fly.

In the real world of FDDI products, most bridges are now SRT-capable, so you'll not likely run into truly incompatible routers. However, you'll undoubtedly have to program SRT routers, and for security reasons you'll want to manually specify the addresses to be forwarded between networks. At that point, you'll have to deal with the kinds of routing—source, transparent, or both—used by stations on your network.

Topology & construction

FDDI is truly a versatile network, in that it supports virtually any physical network topology. You can wire FDDI stations as a physical ring, as satellites off a hub (concentrator in FDDI-speak), or as a tree consisting of combinations of these two topologies, mixing in traditional Token Ring and Ethernet networks (Fig. 6-12). FDDI's high bandwidth makes it practical for full speed disk storage sharing and high performance workstation networking. A key point to keep in mind is that FDDI, as with all fiber networks, uses separate fiber cables for transmitting and receiving. This provides one of FDDI's key mechanisms for fault tolerance, as you'll see shortly.

An FDDI network consists of nodes connected via fiber cables. The nodes are active, as with Token Ring, in that they receive an optical signal, demodulate it to an electrical bit stream, process

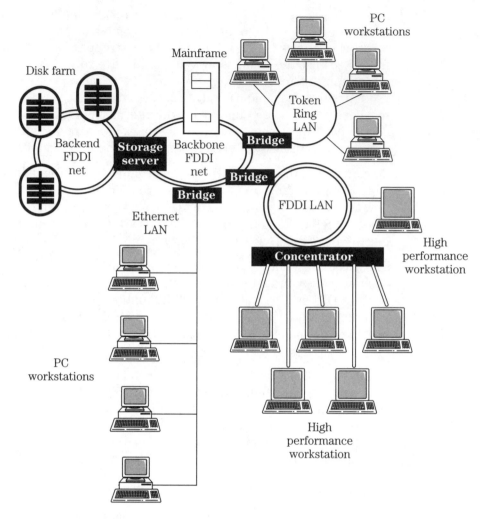

■ **6-12** *FDDI network topologies*

that bit stream, and then remodulate and retransmit the bit stream, possibly modified, on to the next node. FDDI specifies two flavors of node: station and concentrator. A station is an endpoint; for example, a personal computer or mainframe. A concentrator is the equivalent of a Token Ring hub, providing a means to connect stations together.

FDDI has two kinds of stations (Fig. 6-13), and the distinction between them is extremely important. The first kind, called a single-attachment station (SAS), permits connection of a single transmit/receive fiber pair. The second kind, called a dual-attachment station (DAS), has two such pairs, and thus it can connect to two different

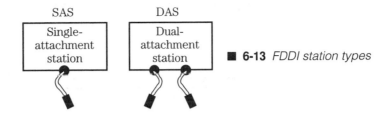

■ **6-13** *FDDI station types*

logical rings. The significance of this is that a DAS lets you build a network having primary and secondary (backup) rings in the same network, which can greatly enhance reliability. DAS stations can be organized as a physical ring. SAS stations, however, must connect to a concentrator, and thus only support a star topology.

FDDI has three kinds of concentrators (Fig. 6-14), which support zero, one, or two attachments. These are called null-attachment (NAC), single-attachment (SAC), and dual-attachment (DAC), respectively. The topological possibilities with FDDI's two kinds of stations and three kinds of concentrators are truly unlimited, as Fig. 6-12 illustrates.

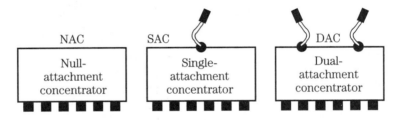

■ **6-14** *FDDI concentrator types*

FDDI's variety of node types is the key to its advanced fault-tolerance features. Chief among these is the ability of an FDDI ring (properly designed) to "heal" itself. Figure 6-15 compares a single-ring concentrator-based SAS network with dual-ring DAS network. Figure 6-16 shows how each topology handles cable failures. In the single-ring network, the concentrator can detect and isolate any failing cable. In the dual-ring network, FDDI automatically reconfigures the ring to continue operation as a single ring, without losing any stations. This self-healing process is called wrapping. Wrapping only protects a dual ring from a single segment fault; a second segment fault would cause the ring to become fragmented, or partitioned. On a partitioned network, however, FDDI SMT can still control the damage so that unaffected stations can continue communicating, a feat not possible with Ethernet, nor with many Token Ring topologies.

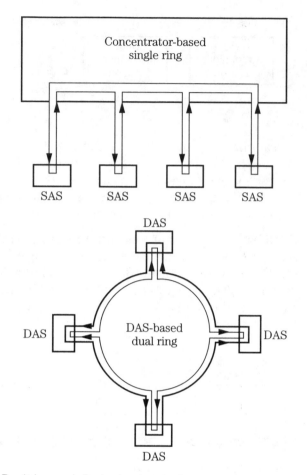

■ **6-15** *Dual-ring and single-ring topologies*

Another FDDI fault-control mechanism is station bypass, the procedure by which a failing or powered-off station (as opposed to a failing segment of cable) is isolated from the network. Token Ring uses this same technique, electrically isolating the failing station by "shorting" the cable across the station. In FDDI, this shorting can also be done optically, using a prism, so that the light signal continues past the bypassed station. In DAS and DAC nodes, the bypassing is done in the station itself. SAS and SAC nodes are bypassed by the concentrator to which they're connected.

Cabling

The medium used in an FDDI LAN depends on the distances to be traversed. In a strictly local network, with segments under 2 km each, you can use multimode fiber and LED-based transceivers.

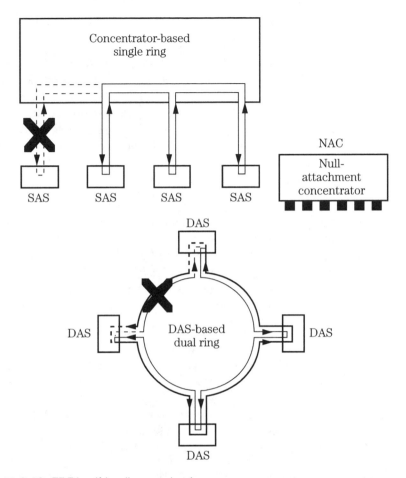

■ **6-16** *FDDI self-healing mechanisms*

For long-haul segments, you'll need single-mode fiber and laser-based transceivers. You can also opt for an alternative medium, called Low-Cost Fiber (LCF). LCF is actually a misnomer, as an LCF network uses the same FDDI-compliant fiber as non-LCF networks. The cost savings comes in the transceivers, which use low power and have higher noise tolerance. The downside is that LCF links are limited to 500 m—but this shouldn't be a problem in intrabuilding installations.

FDDI specifies standard size characteristics for fiber. Single-mode fiber is of size 10/125, which indicates a core of 10 µm and cladding of 125 µm. FDDI calls for multimode fibers of size 62.5/125. You can substitute other sizes (e.g., 50/125, 85/125), but you'll need to perform loss and bandwidth calculations to ensure proper operation. Vendors of fiber cable provide tables and formulas for this purpose. Most vendors also know which of their cable products are FDDI-compliant.

For the most part, fiber installation procedures are similar to those for copper cable. You first select cable with appropriate characteristics for the expected environment: indoor or outdoor sheathing, plenum compatibility, fire resistance, and tensile strength. Cable laid in horizontal raceways sustains less tensile load than cable running through long vertical tunnels, as between floors in a high-rise building. For vertical runs, be sure to consider the weight of the cable when computing tensile loads.

Next, you must ensure that your installation procedures don't overstress the cable. This means using a tensiometer when pulling cable to monitor cable tension, and ensuring that bends are not smaller than the minimum bend radius for the cable, typically 10 to 20 times the outside diameter of the cable. Frequently, a larger conduit contains smaller *innerducts* specifically for fiber cable to reduce friction and ensure that bend radius restrictions are met.

Finally, you need to have on hand the tools necessary for making splices and installing connectors on raw cable ends. Such tools strip off cable sheathing, cleave the fiber to provide an optically clean end, and align the fiber as necessary in the connector or splice. Mechanical splices simply butt the two cleaved ends of fiber against each other, resulting in an unavoidable signal loss. When signal loss isn't tolerable, a special fusion splicer can be used to actually melt the fibers together, resulting in negligible signal degradation.

Copper Distributed Data Interface (CDDI)

If you're incorporating FDDI into an existing network, but you want to extend FDDI's 100-Mbps speed and other architectural advantages to the whole network, you might consider using copper cable for part of the network. The CDDI (Copper Distributed Data Interface) standard specifies the requirements for unshielded twisted pair (UTP), shielded twisted pair (STP), and coaxial (thick and thin Ethernet) network segments. You need to use CDDI, rather than FDDI, station adapters, and at the point where CDDI and FDDI network partitions join, you'll need an FDDI-to-CDDI concentrator. Your existing STP cable plant may well be CDDI-capable. The key requirement is that stations be within 100 m (although some CDDI vendors support up to 150 m).

Not all existing copper cable can sustain CDDI's 100-Mbps data rate. In particular, many kinds of UTP used for telephone wiring and low speed data communications are inadequate for CDDI. UTP cable routing is also critical. A UTP cable rated for 100 Mbps may have been routed too close to electrical noise sources such as ac lines or fluorescent lights. Routing UTP alongside unshielded telephone lines may also disqualify the installed UTP.

To determine if a particular cable can handle CDDI's bit rate, you need to know either the EIA or IBM classification for the cable. These classifications specify number of twists per inch, wire gauge, capacitance, and signal loss of the cable. Only EIA Category 5 and IBM Type 1 cables work with CDDI. If the cable is unshielded (EIA Category 5 is available both shielded and unshielded) you must ascertain that the cable's routing stays at least one foot from all ac electrical sources and does not have any telephone signals carried in the same cable.

Connectors

FDDI's plethora of station types makes for a complex connector situation. FDDI uses a special dual-cable connector to prevent misconnecting send/receive cables, with keying on the connector to make sure the cable can only be connected to appropriate devices. The cables connect to sockets on each device, called ports. Figure 6-17 summarizes FDDI's port and connector standards.

151

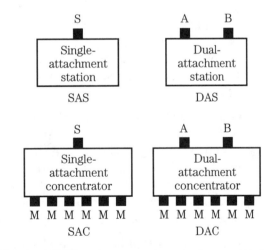

■ **6-17** *Standard FDDI ports, connectors, and connector keying*

Each device has uniquely-labeled ports and the cables correspond to the port names. SAS and SAC devices have a single port labeled S (for *slave*); DAS and DAC devices have two ports, labeled A and B. A complication with DAS/DAC devices is that they support dual rings, with the *in* cable on one connector and the *out* cable on the second connector belonging to one ring, and the reverse pairing belonging to the other. The *A* port is the port having the primary ring's *in* cable and the secondary ring's *out* cable.

Concentrators, of course, also have *spur* ports, to which their subsidiary devices are connected. These ports are labeled *M* (for *master*), and a given concentrator can have any number of them.

We're not done yet. FDDI also makes a distinction between peer and nonpeer ports. Peer ports connect devices at the same or higher levels in topology. Thus the *S* port on an SAC is a peer port, as are the *A* and *B* ports on a DAC. An NAC has no peer ports.

Integrating Token Ring/Ethernet LANs into FDDI

Beyond the physical cabling issues you must face when integrating traditional networks into FDDI, you must consider protocol transport issues. In FDDI's design phase, engineers expected that traditional network protocols, such as TCP/IP and ISO/OSI, would in theory interoperate with FDDI transparently. But early deployments discovered a host of protocol compatibility issues that make transparent interoperability impossible.

IP, for example, uses a data bit order opposite that of FDDI's. FDDI's designers did not consider this a problem, because addresses in IP packets are in canonical form. However, they forgot about IP's Address Resolution Protocol (ARP) and Routing Information Protocol (RIP). Both ARP and RIP carry addresses in the data part of a packet, and those addresses will be in the wrong bit order for FDDI.

There are many such problems, and the only practical solution is to modify higher level protocols to handle differences between FDDI and native protocols. The changes have already been made to the TCP/IP and OSI protocols standards, as well as IEEE's 802 standards. As a network integrator, you'll need to determine that software and hardware you purchase complies with the updated standards for FDDI compatibility.

To summarize, then, FDDI is the natural successor to traditional Token Ring and Ethernet networks. As an ANSI standard (adopted by ISO for worldwide use), it provides a concise specification for a high speed, metropolitan area network. Its optical fiber medium solves most of the shortcomings of copper media: RF interference, crosstalk, corrosion, and security. The light weight and small size

152

of fiber simplify installation and maintenance. All of these features improve overall network reliability.

FDDI was designed with an eye toward easing its incorporation into existing networks. You can deploy FDDI as a backbone connecting lower-speed traditional LANs, or merge existing LANs into larger FDDI LANs using CDDI to combine fiber and copper media.

By providing for synchronous, time-critical services, FDDI makes possible the integration of voice and multimedia applications on existing networks. The FDDI-II standard enhances this capability by supporting isochronous services.

FDDI's wide variety of node types makes possible an unlimited range of network topologies, combining rings and trees as required to meet cost and reliability requirements.

FDDI's fault tolerance makes it a good choice for mission-critical network design, and its steadily declining costs make FDDI-to-the-desktop practically inevitable. FDDI's distributed station management make fault recovery automatic.

Finally, FDDI has a future. Beyond the FDDI-II standard, which is essentially complete, the FDDI Follow-On LAN (FFOL) promises 1 gigabit per second speeds. FDDI can also serve as a local transport medium in new-technology MANs and WANs based on ATM or SMDS. The fact that FDDI's developers have taken care to provide a growth path makes it a good next-generation replacement for existing LANs.

Frame & cell relay

Introduction to frame & cell relay

High speed packet switching technology, or fastpacket, evolved out of the need for high-bandwidth-on-demand wide area networking. LAN traffic is changing as data volume increases and new applications appear. Traffic has become more "bursty," with higher peaks and lower valleys in utilization patterns. New applications tend to move large volumes of data at infrequent intervals, and that data needs to move quickly to maintain application responsiveness. The once-apocryphal example of a doctor reading CAT scan images remotely is now an everyday occurrence. What's more, end users want the same technology for more mundane purposes, such as processing insurance claims and archiving routine paper documents.

And why shouldn't they get it? After all, fiber optic channels are capable of virtually unlimited bandwidths. Why can't we just run fiber cables everywhere? If you've looked into running your own FDDI fiber lines cross country, you already know the answer: it would simply cost too much. The cost of the medium—copper or fiber—is insignificant compared to the cost of installation. Yes, the current long distance network is almost entirely fiber. But that doesn't deliver bandwidth to end users, who live out on the telephone network's euphemistic "last mile," the copper loops connecting customers to their central office (CO). This is where most of the cost of the telephone network lies, and replacing it all with fiber would cost, by conservative Bellcore estimates, some $250 billion and twenty years. Yes, eventually fiber will replace the last mile, but not in the useful future.

So, although we have the technology, we don't have the pocketbook. This makes existing broadband fiber channels in the long distance network precious commodities. Such commodities can't be wasted by dedicating circuits to customers who use the available bandwidth only sporadically. Fastpacket services make efficient use of this constrained resource while giving customers high bandwidth when needed. Fastpacket will carry the day until Fiber-to-the-Home (FTTH) arrives sometime around the year 2020.

Fastpacket technologies

Two of the most important fastpacket technologies are Frame Relay Service (FRS) and Cell Relay Service (CRS). Their names come from the idea of relaying packets quickly from one station to another in a wide area network. Think of runners passing a baton in a relay race.

FRS provides speeds from 56 kbps to 1.544 Mbps; CRS from 56 kbps to 622 Mbps. Both have well-defined implementation standards: ANSI T1.606/617/618 for FRS, and ITU.121/2 for CRS. Both services were primarily designed for metropolitan areas, although both will eventually work over arbitrary distances, as the underlying long-distance infrastructure gradually moves to faster optical links.

The concept of fastpacket services is simple: instead of providing customers with dedicated-bandwidth bit pipes, provide instead a variable-bandwidth packet-switching service. Figure 7-1 illustrates a WAN using dedicated digital services to connect four LANs. The WAN requires separate lines between all network entities to ensure coverage—in this case, three lines from each LAN site. These dedicated bit pipes (e.g., T1 lines, discussed in detail in Chapter 8), are not always running at capacity, wasting valuable bandwidth. The wasted bandwidth costs both the customer and the supplier money: the customer must pay for bandwidth even when it isn't in use, and the supplier must build infrastructure far in excess of that actually carrying data. LAN traffic by nature tends to be bursty, requiring high bandwidth infrequently, but some bandwidth nearly all the time, so the connections must be available full-time. And while dedicated digital networks are capable of incremental bandwidth-on-demand changes, such changes are not instantaneous, and persist beyond the time required for the increased traffic. More and more, incremental bandwidth adjustment is too slow and too coarse-grained to respond to traffic changes quickly or efficiently enough. The dedicated-line WAN also requires separate physical attachment hardware (e.g., CSU/DSU) for each line, further increasing costs.

With FRS and CRS, each site has a single connection to the fastpacket service (Fig. 7-2). The single connection requires only a single physical attachment unit, and any site can route packets to any other site via the fastpacket network. The public fastpacket

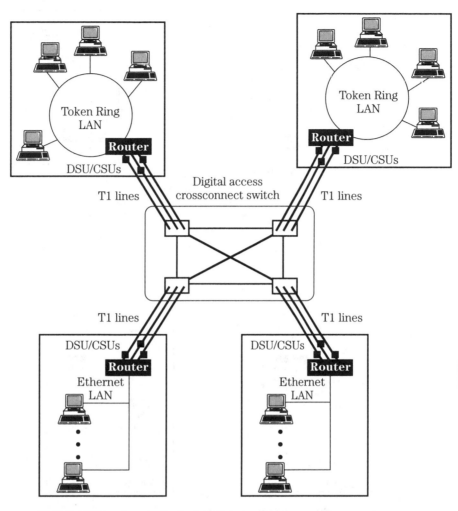

■ **7-1** *WAN using dedicated digital circuits*

network is being shared with many other customers, all of whom need bandwidth-on-demand. The fastpacket service interleaves the packets from all users without permanently setting aside bandwidth for any one. Statistically, higher bandwidths won't be required simultaneously for all users—traffic will be high for some users some of the time, but not all users all of the time. This averaging of bandwidth requirements makes more efficient use of the broadband facilities inside the network.

With either FRS or CRS, you contract with a supplier for a certain guaranteed bandwidth, with allowances for high-volume bursts; the supplier bills you a monthly fee plus possibly usage charges. You're guaranteed that your average throughput will equal the con-

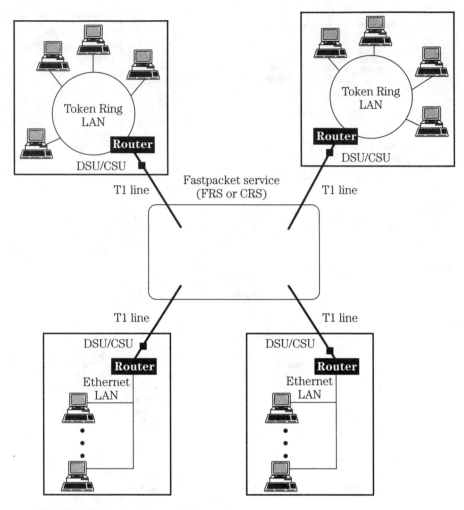

■ **7-2** *WAN using fastpacket services*

tracted bandwidth, and that bursts up to a certain amount will always be serviced. Because the supplier saves money on the physical plant by sharing wideband facilities among many users, the cost per bit moved is much lower than the costs using equivalent dedicated facilities. Thus the supplier can lower its rates, and precious wideband facilities aren't frittered away. A significant advantage of fastpacket services is that they are distance-insensitive. That is, you don't pay mileage or long-distance fees to transport data—all users connected to the same fastpacket "cloud" can interoperate at a fixed cost, and the cloud can span hundreds of miles.

Both FRS and CRS require special switching technology at the central office, which means that service providers can't use their existing voice digital switches. Fastpacket uses special, dedicated fastpacket switches. FRS switches are often proprietary, and thus don't always interoperate with networks using switches by different manufacturers. CRS, however, uses Asynchronous Transfer Mode (ATM) switches, which have an international standard supporting interoperability. ATM switches can be used alongside voice switches, and eventually replace them.

FRS and CRS differ in a few other important ways. FRS appeared first on the market, and is a data-only service. It uses existing digital voice broadband transport facilities, T1 and T3 (DS1 and DS3), both for customer-to-switch links and inter-switch links—hence its maximum rate of 1.544 Mbps, which derives from T1's 1.544 Mbps rate. CRS came later, supports integrated voice and data, and requires ATM technology. ATM itself isn't married to T1/T3 transport facilities, although it can use them. Instead, ATM was designed to use the new broadband backbone called SONET (Synchronous Optical Network), which is gradually being deployed in the long-distance network. As SONET becomes more widely available, ATM will automatically take advantage of it. In the meantime, ATM moves data within a switch at 1200 Mbps, fast enough for many channels of broadcast video service, and between switches at whatever broadband speeds are available. ATM offers two service speeds: 155 Mbps and 622 Mbps.

Customers connect to either FRS or CRS at the network interface, using a physical access channel (e.g., T1, T3), called the local transport facility. Unlike the fastpacket switching service itself, the local transport is a dedicated line, and must thus support bandwidth up to the maximum possible burst rate. This bandwidth is called the aggregate data rate. In addition to T1 and T3 local transport, fastpacket also supports fiber local transport, in the form of SONET connections.

Because FRS by itself doesn't support voice, it isn't part of the group of standards collectively called *Broadband Integrated Services Digital Network* (B-ISDN). ATM, however, *is* a B-ISDN standard, and although ATM's CRS was designed to replace FRS, ATM can "grandfather in" FRS to preserve end user equipment investments. This turns out to be useful for more than just backward compatibility. FRS also provides a useful *local* transport—from the customer to the public ATM switch—for data rates below the FRS maximum of 1.544 Mbps. FRS uses only T1 copper lines for local transport. When CRS local transport isn't available, FRS can span the gap for many applications.

When CRS local transport is available, it supports 64n kbps aggregate bandwidth increments, where n ranges from 2 to 672, which translates to network interface speeds of 128 kbps to 43.008 Mbps. CRS can use T1 or T3 copper-based, or SONET fiber-based local transport.

FRS is much more widely deployed than ATM, but that will change as more ATM switches are installed. The existing plant of FRS-only switches will probably not grow significantly, because an ATM switch provides both voice and data capabilities for little additional cost. More importantly, though, ATM is readily deployable as a LAN in its own right. This means that a company can install an ATM switch on-site as a network backbone, or even as the primary LAN itself. ATM switch prices are dropping steadily, running as low as $20,000 to $50,000 for low-end products. ATM-compatible bridges, routers, and workstation attachments still cost several times as much as non-ATM components. Even now, though, these costs are dropping enough to bring ATM directly to the desktop.

Fastpacket fundamentals

160

A packet-switched data service accepts packets from an end user application, transports them through a network of computers, and delivers them to a specific end user application. The advantages of packet switching are the effective sharing of local transport and wideband carrier facilities, and shorter connection setup times. The disadvantage is the variable delay that packets experience en route to their destination.

The traditional public packet switched data network (PSDN) uses X.25 as its underlying protocol, and supports network interface speeds of only 56 kbps. X.25 guarantees reliable transport, which means that every packet is acknowledged along its entire route, and any errors are corrected by the network retransmitting the packet (Fig. 7-3). Each station along the route stores a packet until it gets a positive acknowledgment from the adjacent station. This imposes considerable protocol overhead for error control information. When X.25 was designed, its error detection and correction process was deemed necessary given the low reliability of the old copper-and-microwave long distance networks of the time. It is such fastidious error control that accounts for X.25's low data rate, which is too slow for most LAN interconnection applications.

Fastpacket technology offers higher throughput by providing faster network interface speeds (many times X.25's 56 kbps limit)

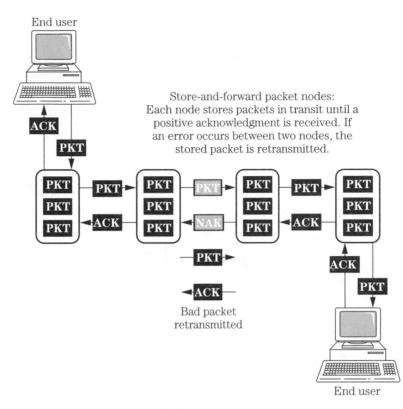

End user

Store-and-forward packet nodes:
Each node stores packets in transit until a
positive acknowledgment is received. If
an error occurs between two nodes, the
stored packet is retransmitted.

Bad packet
retransmitted

End user

■ **7-3** *Traditional packet switching network (X.25)*

and by eliminating packet overhead for error correction. Higher network interface speeds come from fast digital local transport (e.g., T1). Error correction can be eliminated because the inherent reliability of the current digital carrier network, where error rates are low, makes packet transmission errors an infrequent phenomenon. Rather than saddle the low-level data link layer with error control that is seldom invoked, fastpacket protocols simply discard corrupt packets when they occur, leaving higher-level protocols to handle error detection and retransmission (Fig. 7-4). Higher-protocol error recovery is slower than recovery in the data link itself, but because errors are infrequent the increased delay is insignificant. In the meantime, packet routing is faster, because nodes need not store packets for possible retransmission, and decreased protocol overhead reduces the amount of unproductive data traffic.

All packet switching uses the concept of a *virtual circuit* (VC). The idea is straightforward: each packet has a header associated with it containing a *virtual circuit identifier* (VCI) associating

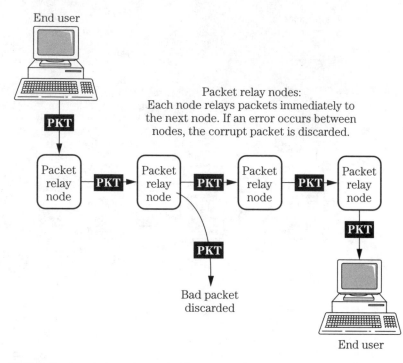

End user

Packet relay nodes:
Each node relays packets immediately to
the next node. If an error occurs between
nodes, the corrupt packet is discarded.

PKT

| Packet relay node | **PKT** | Packet relay node | **PKT** | Packet relay node | **PKT** | Packet relay node |

PKT

PKT

Bad packet
discarded

End user

■ **7-4** *Fastpacket packet switching network*

the packet with an established logical connection, or VC (Fig. 7-5). The header is added to the packet by a network device (sometimes implemented as host-resident software) called the *Packet Assembler/Disassembler* (PAD). Many VCs can exist over a single physical connection, up to the maximum numerical value of the VC field in the packet header. X.25 supports up to 256 virtual circuits, FRS up to 1024, and CRS up to 65,536. VCs can be permanent (PVC), with the virtual connection (and route through the packet switch) established once by the service provider, or switched (SVC), established via a call setup procedure by the end user every time the circuit is used. PVCs are like leased lines, and have fixed VCIs assigned at installation time. SVCs get a new unique VCI whenever a new connection is requested; they're similar in concept to switched dial-up lines. Note that VCIs have local significance only—the VCI on the other end of a VC will likely have a different value. This isn't a problem though, as VCIs are simply for keeping VCs separate at the network interface. The internal network will use different VCIs for each hop in the route.

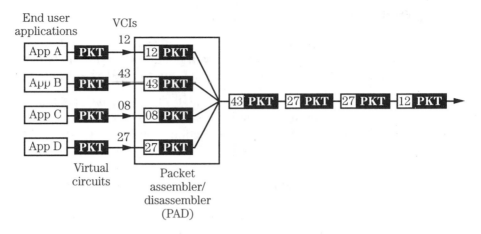

End user applications

VCIs

Virtual circuits

Packet assembler/ disassembler (PAD)

■ **7-5** *Virtual circuits in a packet switching network*

Frame Relay Service

As noted earlier, FRS is a data-only service. It's concerned only with bursty data traffic, with no provisions made for time-sensitive real-time traffic such as voice or video. Two important terms to know as a consumer of FRS are *committed information rate* (CIR), the guaranteed average data rate for which you've contracted, and *committed burst size* (CBS, also denoted by B_c), the maximum number of bits that can be transferred during some time interval T. The relation between these quantities is

$$T = \frac{B_c}{CIR}$$

For example, given a CIR of 128 kbps and a CBS of 512 kilobits, T is 512 kilobits divided by 128 kbps, or 4 seconds. This means the network guarantees to move 512 kilobits over any four-second period. When buying FRS, you'll need to carefully select CIR and CBS values that yield T values large enough to cover your worst-case bursts. However, another factor enters into the equation, letting you move data in excess of the CBS. That factor is called *excess burst size* (EBS, also denoted by B_e). If the network is congested, you're guaranteed to get performance matching your CIR and CBS. On an uncongested network, though, you can move up to $B_c + B_e$ bytes per second. In the above example, with CBS of 512 kbps, an EBS of 256 kb lets you move data at 768 kb when the network is not congested.

163

Architecture

The heart of FRS is the packet being switched, called, unsurprisingly, a frame. Each frame has a fixed header and a variable-size payload (Fig. 7-6). Here is a quick rundown of each frame field:

- ☐ Flag. Indicates the start of the frame. A similar flag following the last frame field indicates the frame end.
- ☐ DLCI. The Data Link Connection Identifier, a ten-bit field (note that the bit field is split) identifying the VC to which the frame belongs.
- ☐ C/R. The command/response flag, used for local transport flow control.
- ☐ EA. Two separate address extension bits that permit expansion of the DLCI within the carrier network.
- ☐ Congestion Information Bits. A set of flags recording network congestion encountered while a frame is en route. These flags are described below.
- ☐ Payload. A variable-size byte stream, from 262 to 8000 bytes (the maximum size is determined by the service provider) with a prepended length indicator.
- ☐ Frame Check Sequence (CRC). A checksum used to validate the frame while en route.
- ☐ Flag. The trailing flag indicating the end of the frame.

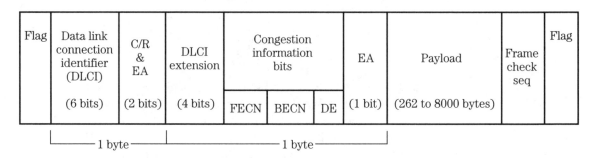

■ **7-6** *FRS frame format*

Note that unlike LAN packets, the frame doesn't contain source or destination addresses. That's because the source and destination addresses are specified for the connection at installation time (for PVCs) or during call setup (for SVCs). In either case, the result is a DLCI identifying the VC associated with that connection.

The Frame Check Sequence is computed when a frame is created and injected at the network interface. It's checked after each hop in the FRS network, and if the check fails the frame is dropped. As described earlier, a higher-level protocol will eventually notice the missing packet and initiate some sort of error recovery.

Congestion control

To meet customers' data-rate requirements, the Congestion Information field records congestion problems encountered while a frame is en route. The algorithm for handling congestion is similar to that used by airlines when they overbook flights. When too many passengers show up for a flight, the airline asks for volunteers to "bump" to a later flight, offering some sort of compensation (savvy travelers always hold out for a free ticket). Usually enough volunteers can be found to avoid inconveniencing priority passengers. In rare instances, "unwilling" volunteers must be chosen by the airline to get the passenger count down to the aircraft's capacity, which understandably results in passenger complaints.

The Congestion Information field contains a *discard eligibility* (DE) flag, which is set in frames "volunteering" to be sacrificed in the event of an overload. The DE flag for a given packet gets set whenever the delivered data rate inside the network exceeds the subscriber's CIR. Such frames are by definition part of a high speed burst, and are considered to have lower priority than other frames. End user equipment can also set the DE flag if it knows a particular frame is nonessential (e.g., certain network management messages).

The network keeps track of congestion problems by setting one of the two explicit congestion notification bits: *forward explicit congestion notification* (FECN), and *backward explicit congestion notification* (BECN). These bits advise the receiving and transmitting ends of a connection, respectively, to throttle back frame traffic rates. Because explicit congestion notification is advisory only, it can be ignored. This is like the airline overbooking scenario where not enough passengers volunteer to be bumped. When this happens, frames are simply discarded if they can't be moved forward in the network. Higher-level protocols must detect and recover from the problem. Like the unwillingly bumped airline passenger, higher-level protocols may complain of excessive error rates. And as with the disgruntled passenger, the remedy is the same: raise Cain with the carrier to get better service. In both cases, the carrier's correct response is to add additional facilities (aircraft seats or packet switching capacity).

Interconnecting LANs using Frame Relay Service

FRS has a lot going for it as a LAN interconnection technology. First, the traditional advantages of packet switching accrue to FRS: a single physical network connection cuts hardware and line costs, bandwidth-on-demand supports bursty traffic patterns, and charges are incurred only when data moves.

Second, the variable-size information frame can accommodate virtually any kind of embedded LAN packet (Fig. 7-7). This greatly simplifies the design of FRS-capable bridges and routers, which can simply wrap an FRS frame around the LAN packet and ship it across an appropriate VC.

Header	Payload (up to 8000 bytes)	Trailer

X.25 packet

TCP/IP logical link control

SNA logical link control

Ethernet media access control

Token Ring media access control

■ **7-7** *FRS frames easily carry embedded protocols*

Third, high burst data rates enable WAN-connected LANs to operate at near-local performance levels, making remote access to new, high-bandwidth applications practical.

Finally, FRS's long-distance insensitivity makes possible LAN interconnects across large metropolitan areas with uniform costs per node. As long as all the nodes connect to the same FRS cloud, no inter-exchange carriers get involved and the line costs are based purely on bandwidth.

The primary considerations for FRS/LAN interoperability are maximum frame payload and CIR/CBS values. You should ensure that the maximum payload supported by your FRS supplier accommodates the largest packets in the LANs you plan to interconnect. Otherwise packet fragmentation will greatly slow throughput and may even cause some protocols to fail (e.g., TCP/IP Address Resolution Protocol).

You should select a CIR value with some margin for error, after measuring the actual traffic rates on the LANs in question. Thus if network analysis shows an average traffic flow of 220 kbps, a CIR of 256 kbps will help prevent congestion rejection if traffic occasionally exceeds this average. Some FRS providers permit arbitrary CIR values; others permit only fixed-size increments (e.g., 64 kbps).

Similarly, you should choose a conservative CBS value. If you measure peak bursts of 900 kilobits over two- to three-second intervals, a CBS value of 1000 kilobits will ensure that slight traffic increases don't congest your FRS link. Some FRS suppliers sell only specific CIR/CBS combinations (e.g., 64 kbps and 256 kilobits, 256 kbps and 768 kilobits, etc.). Similarly, the EBS value is tied to some fixed fraction of the CBS value (e.g., one-half).

You'll have to choose a local transport service—the channel that connects your network interface to the FRS switch—from available digital link speeds provided by your local exchange carrier (LEC) (digital links are described in more detail in Chapter 8). The digital link must have enough capacity to handle the highest peak traffic, but because such links come in fixed increments, you'll have to buy the next-higher capacity service. For example, if you measure the burst size on a LAN at 1.024 Mbps, expect to lease a 1.544 Mbps T1 link if your LEC doesn't have fractional T1.

Figure 7-8 illustrates an example WAN without and with FRS. In this scenario, a manufacturer has plants in various cities interlinked to form a WAN. In the dedicated circuit implementation (Fig. 7-8a), each link is sized for the maximum burst rates along each possible LAN connection path, using leased T1 fractions ranging from 256 kbps to 1.5 Mbps. The FRS solution (Fig. 7-8b) uses permanent virtual connections (PVCs) and is sized with much lower-capacity CIRs of 64 kbps and appropriately selected CBS values for each node. FRS ends up costing about 40% less than the dedicated circuit plan.

ATM & Cell Relay Service

ATM and Cell Relay Service (CRS) are father and son technologies. As part of the Broadband Integrated Services Digital Network (B-ISDN) blueprint, ATM supports data, voice, image, and video.

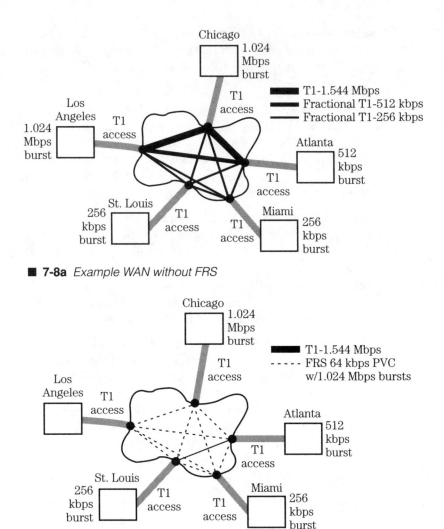

■ 7-8a *Example WAN without FRS*

■ **7-8b** *Example WAN with FRS*

CRS is an end user interface to ATM. Like FRS, ATM and CRS are packet switching technologies. Unlike FRS, though, ATM/CRS uses fixed-length frames, called *cells*, to carry all data. A cell is 53 bytes long (5 bytes header and 48 bytes payload); its constant size makes routing of packets even faster than FRS switching. Fixed-size cells also let ATM guarantee on-time delivery of voice and video data. The asynchronous in ATM's name refers to its ability to transfer data on separate virtual circuits at variable bandwidths. Contrast this with the Synchronous Optical Network (SONET), which moves data over channels having dedicated bandwidth, using time division multiplexing (TDM).

ATM requires special switching systems, which can either be part of a public network, the hub of a corporate LAN, or a workgroup hub for a small cluster of workstations. This is a significant feature, as no other fastpacket technology can scale from LAN to MAN to WAN sizes. ATM switches for a few hundred nodes cost under $100,000, well within the budgets of medium-sized companies. Workgroup-size ATM switches can be had for around $25,000, and support up to 50 users. ATM interface cards for individual workstations cost as little as $500 to $1000, putting ATM-to-the-desktop on the near horizon. What's more, ATM can serve as a corporate voice telephone switch, making justification of private ATM networks even easier.

As an umbrella standard, ATM supports several service classifications, as shown in Fig. 7-9. The classifications and their associated characteristics are:

Class A: Timing required, constant bit rate, connection oriented
Class B: Timing required, variable bit rate, connection oriented
Class C: Timing not required, variable bit rate, connection oriented
Class D: Timing not required, variable bit rate, connectionless
Class X: Unrestricted (variable bit rate, connection oriented or connectionless)

Attributes	Service classes				
	Class A	Class B	Class C	Class D	Class X
End-to-end timing	Required		Not required		
Bit rate	Constant	Variable			
Connection mode	Connection-oriented			Connection-less	Either
Services	Voice and video	Un-defined	FRS	SMDS	CRS

■ **7-9** *ATM service classes*

Class A provides the constant bit rates and guaranteed timing required for voice and video applications. Class B isn't currently defined for any particular application, but its characteristics make it useful for compressed video distribution (e.g., delivering movies to home subscribers for storage and offline playback). Class C is used for "grandfathering" Frame Relay Service (FRS), and Class D provides a grandfathering mechanism for Switched Multimegabit Data Service (SMDS). Class X carries Cell Relay Service (CRS), and is the only class discussed in detail in this book.

Architecture

Both ATM and CRS (which is simply the end user interface to ATM's internal cell transport mechanism) use a 53-byte cell (Fig. 7-10) with the following fields:

- Generic Flow Control (GFC). A four-bit set of flags used to provide local flow control. The GFC is ignored by the network.
- Virtual Path Identifier/Virtual Channel Identifier (VPI/VCI). A 24-bit virtual connection number. Not all bits are used; most ATM equipment supports 4096 unique VPI/VCI values.
- Payload Type Indicator. A three-bit code point marking a packet as user data (000 through 011) or management data (100 through 111).
- Cell Loss Priority. A one-bit flag letting end users assign higher reliability to certain packets. Currently not slated for actual implementation (note that ATM does not perform data error detection).
- Header Error Control. An eight-bit check byte for performing header error detection.

GFC	Virtual path identifier (VPI)	Virtual path identifier (VPI)	PTI	CLP	Header error control (HEC)	Payload
(4 bits)	(8 bits)	(16 bits)	(3 bits)		(8 bits)	(48 bytes)

└─ 1 byte ─┴─ 1 byte ─┴─ 1 byte ─┴─ 1 byte ─┴─ 1 byte ─┴──── 48 bytes ────

■ **7-10** *ATM/CRS cell format*

The ATM/CRS header has few similarities to FRS headers. ATM's 24-bit VPI/VCI is much larger than FRS's 10-bit DLCI; the larger value gives ATM plenty of expandability and the future capability to address many, many local entities. Eventually, ATM may extend network intelligence to all sorts of mundane business and personal electronic devices: telephones, fax machines, copiers, televisions, and even toasters (well, maybe not toasters).

ATM's cell header also lacks the congestion control and CRC checksum of the FRS header. In ATM, congestion algorithms and signaling operate out-of-band, transparently to the end user equipment and higher-level protocols. ATM is even more cavalier about data errors: it doesn't detect them at all. Instead it cheerfully delivers corrupt packets to the higher-level protocols, which are responsible for noticing an error (through checksums in embedded packets) and performing error recovery. ATM does perform error checking on the cell header itself, to avoid delivering cells to incorrect destinations.

ATM cells are, by prearrangement, identical in size and format to Switched Multimegabit Data Service (SMDS) cells, which ensures interoperability with this proprietary Bellcore service (discussed in detail in Chapter 9).

Like FRS, ATM's architecture supports both PVC and SVC connections, although only PVC connections are being offered in initial ATM deployments. With both PVCs and SVCs, a connection can specify certain useful operating parameters:

☐ The identity of the calling and called parties.
☐ A requested cell transfer rate, in cells per second. The rates for each direction can be specified separately, permitting asymmetric connections (e.g., for file transfer from server to client).
☐ A quality of service class (high or low, specifying either very reliable or best-effort delivery).
☐ Routing and VPI/VCI preferences.

A PVC requires setting these values at installation. An SVC sets the values at call-setup time, and also permits changing the values while a connection is in progress, making for more dynamic bandwidth management on the customer end than FRS supports.

Overall, ATM and CRS offer a superset of FRS capabilities. This lets FRS and CRS interoperate easily. For example, you might use FRS to connect to a CRS service when constructing a corporate MAN backbone. Unlike FRS, you generally don't have to specify specific

171

CIR/CBS values when contracting for CRS. ATM's much broader internal bandwidths can absorb a wider range of load conditions, eliminating the need to enforce bandwidth constraints. You will, to be sure, get charged for the bandwidth you actually do use.

ATM as a wide area solution

Figure 7-11 shows ATM used in a publicly provided MAN service to interconnect several corporate LANs within a single city. In this scenario, CRS provides a high speed backbone service for corporate MAN.

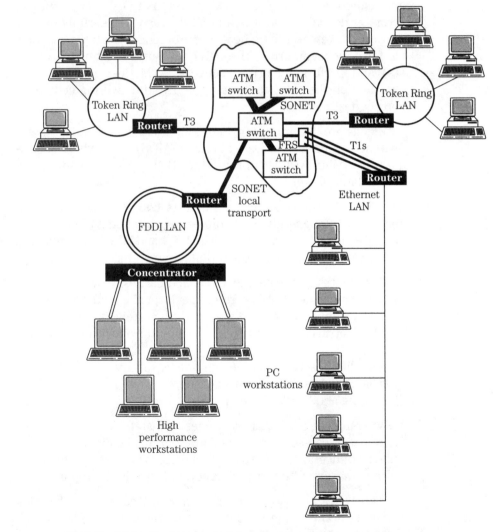

■ **7-11** *ATM in a wide area environment*

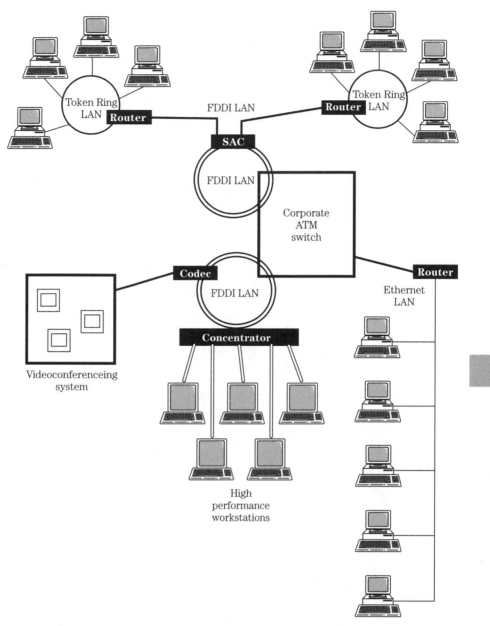

Token Ring LAN

Router

FDDI LAN

Token Ring LAN

Router

SAC

FDDI LAN

Corporate ATM switch

Codec

FDDI LAN

Router

Ethernet LAN

Videoconferenceing system

Concentrator

High performance workstations

■ **7-12** *ATM in a local area environment*

173

The network provider's CRS network consists of a single ATM switch, or a small number of interconnected ATM switches. Each corporate LAN connects to the CRS via a CRS-capable router. These routers connect to the ATM switch over available LEC transport services, including T1 and dedicated fiber.

The network is configured to set up multiple virtual channel connections among all of the corporate subnets, ensuring that any station on any network has full LAN bandwidth connectivity across the MAN.

Today such a backbone is limited to a single geographical area. As SONET becomes more widely deployed within the long distance network, network providers will be able to interconnect ATM switches.

ATM as a local area solution

Figure 7-12 on the previous page shows ATM used as a private corporate LAN backbone. In this scenario, an in-house ATM switch interconnects FDDI, 802.3/Ethernet, and 802.5/Token Ring LANs using FDDI between the switch and ATM-compatible routers. The broadband FDDI links can also connect natively attached high-performance workstations, video-conferencing systems, and other ATM-capable devices.

174

V

Traditional WAN interconnections

WITHOUT A DOUBT, THE FASTPACKET TECHNOLOGIES AND services described in Chapter 7 are the future of LAN interconnect. Unfortunately, the future isn't here yet. Frame Relay Service is just now being deployed in a few major metropolitan areas, and Cell Relay Service, along with its ATM technology umbrella, is used primarily in private networks.

Until these services are widely deployed, the currently available suite of "traditional" WAN interconnections will remain common. These services include both point-to-point and multipoint links.

In the point-to-point department, you can choose from among the always-available (and always slow) dial-up modem connection, leased analog lines, and switched and leased digital services of various kinds. Point-to-point links provide guaranteed bandwidth and availability, which is important when you have consistently high LAN traffic or when you're mixing network connections with other telecommunications services such as digital voice and teleconferencing. Point-to-point data rates for modems range up to 28.8 kbps, and for other digital connections, up to 45 Mbps.

When LAN traffic interconnects need not be full-time, you can often save money by using multipoint services. These have the advantage of letting you establish connections only when needed, and only using the amount of bandwidth needed at any given time. Multipoint interconnects include the well-established X.25 Public Switched Packet Network (PSPN) for data rates of up to 56 kbps,

Integrated Services Digital Network (ISDN) for rates from 56 kbps to 1.5 Mbps, and Switched Multimegabit Data Service for rates of 1.544 Mbps to 45 Mbps.

History of digital services

All of these services utilize telephone company wide area digital facilities. These facilities were originally developed by Bell Telephone to carry voice traffic between central offices and long distance switching centers. To understand how today's digital services came about, you need to understand why and how the long distance voice network "went digital" in the first place.

Prior to digital long-distance facilities, transmitting analog voice signals over long distances required expensive signal regeneration equipment (Fig. V-1). Because an analog signal can never be regenerated perfectly, the repeated regeneration required for long distance calls resulted in increasingly degraded sound quality in direct proportion to the distance traveled.

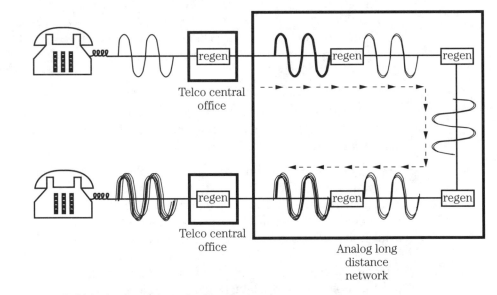

■ **V-1** *Analog voice telephone network*

By digitizing voice calls at the central office, the calls could be treated as data, using digital routing and signal regeneration (Fig. V-2). Digital error detection and correction techniques mean that a digitized voice signal can be regenerated perfectly. When the digitized voice is converted back to an analog signal at the destination, it is as clear as it was when it arrived at the CO from the

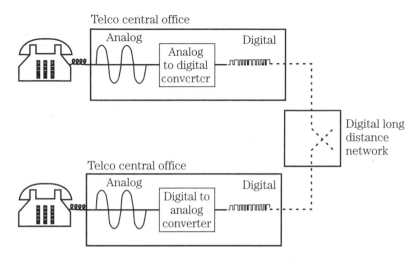

■ **V-2** *Digital voice telephone network*

subscriber. The only opportunities for degradation occur in the local loop, between the calling parties and their respective COs.

The technique for converting voice signals to digital data is called Pulse Code Modulation (PCM) (Fig. V-3). The process is simple: the amplitude of the audio signal is sampled many times per second by an analog-to-digital converter, with each sample being converted into a number representing the signal amplitude at that instant. The stream of numbers resulting from this digitization process is the digital data stream transmitted by the network. At the destination, a digital-to-analog converter generates an audio signal from the digital data stream that closely approximates the originator's audio signal.

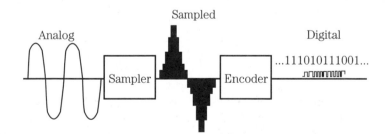

■ **V-3** *Pulse Code Modulation*

You'll want to pay attention closely now, because here is where the foundation of the current digital network is laid. The standard for PCM calls for a sampling rate of 8000 samples per second, with each sample ranging in value from 0 to 255, which can be represented as an 8-bit binary number. 8000 samples per second times 8 bits per sample yields a data rate of 64,000 bits per second, or 64 kbps. Hmmmm, where have you heard that number before? It sounds suspiciously like the basic data-rate building block of many traditional data services, doesn't it? It should come as no surprise, then, that when telcos began offering digital service, they did so at data rates in multiples of 64 kbps.

The long distance network consists of a hierarchy of Digital Signal rates (Fig. V-4). The lowest rate, 64 kbps, is called DS0. The next higher rate, DS1, is 24 times faster at 1.544 Mbps. The remaining two rates, DS2 and DS3, are 6.176 Mbps and 44.736 Mbps, respectively. The physical facilities used to carry multiple DS0 signals are called trunk lines, and have names corresponding to their DS rates: T1 for DS1, T2 for DS2, and T3 for DS3. In Europe the bundling is different, with a trunk containing 32 instead of 24 DS0 channels. That facility is called an E1 line; it's not discussed further in this book. Neither is the T2 line, which while part of the hierarchy is not used in practice.

Designator	Data rate	Trunk facility
DS0	64 kbps	Non-basic rate
DS1	1.544 Mbps	T1
DS2	6.176 Mbps	T2
DS3	44.776 Mbps	T3

■ **V-4** *Digital Signal (DS) rates*

Originally, Bell Telephone used its digital facilities strictly for internal routing of telephone calls. But with divestiture and open competition in the long distance market, telcos decided to sell their digital facilities directly to end users. The various services that deliver access to these facilities all rely on the underlying DS carrier structure, which makes them both easy for telcos to deploy and somewhat difficult for end users to use efficiently.

A company using digital data services to link WANs must buy enough dedicated capacity to meet peak needs, which is frequently far in excess of the average traffic rate. For low data rates (56 kbps and below), the X.25 PSPN offers an attractive bandwidth-on-demand solution. For higher rates, though, digital data services offer only limited bandwidth-on-demand solutions. Of the available options, ISDN has perhaps the most flexibility. Unfortunately, it is one of the least readily available services today. No matter what traditional digital service you choose, you won't be able to both maximize efficiency and minimize cost; attaining one of these goals always compromises the other.

Point-to-point links

Introduction to point-to-point links

Point-to-point links establish a physical connection between a local and remote station. These links come in a variety of data rates, and as speed and capability increase, cost, not surprisingly, does also. Because a point-to-point link provides *dedicated bandwidth* for the duration of the connection, the cost of moving data this way is usually much higher than with fastpacket services (described in Chapter 7). Also, when constructing a MAN or WAN using point-to-point links, you must buy these dedicated facilities for each line of communication you want to establish. This means that the number of links increases rapidly with the number of nodes: three links for three nodes, ten links for five nodes, and so on. Fastpacket services let you establish the necessary number of dedicated links as *virtual circuits* over a shared communications service; but until ATM, Frame Relay Service, and Cell Relay Service become fully deployed, switched circuits will be one of the WAN solutions in your tool kit.

Local telephone companies, also called *Local Exchange Carriers* (LEC), offer traditional telecommunications services in an ascending ladder of capabilities and costs (Fig. 8-1). These services use existing telco voice network facilities and the copper local loop between the customer and central office. The costs shown in Fig. 8-1 are approximations, and recently some LECs have been drastically cutting them. For example, Pacific Bell in California now offers T1 circuits for as little as $350/month plus distance charges. The moral: check prices carefully and frequently.

The bottom rung of the point-to-point ladder is the analog connection, using modems to carry data over leased or switched lines. Leased lines are full-time connections between two specified locations; switched lines are regular phone lines (often called POTS, for Plain Old Telephone Services). Modems have become extremely popular in the first half of the '90s, but they've reached their maximum speed limit due to the dictates of physical law. The

Service	Line speeds	Required equipment	Approximate costs	
			Line	Equipment
Analog, leased, and switched	100 bps– 28.8 kbps	Modem (2)	$ 20/mo– 150/mo distance based	$ 100– 1000
Digital Data Service (DDS)	2400 bps– 56 kbps	DSU/CSU (2)	$ 50/mo– 500/mo distance based	$ 100– 1000
Switched-56	56 kbps	DSU/CSU (2)	$ 50/mo– 250/mo	$ 300– 1000
Fractional T1	64 kbps– 1.544 Mbps	DSU/CSU (1) or T1 mux	$ 100/mo– 2000/mo	$ 500– 5000
T1	1.544 Mbps	T1 Multiplexer (2)	$350/mo– 2000/mo	$ 1000– 5000
T3	44.736 Mbps	T3 Multiplexer (2)	$2500/mo– 10,000/mo	$ 1000– 15,000

■ **8-1** *LEC point-to-point services*

fastest modems today, the so-called V.34 or V.fast standard, oper-ate at an uncompressed rate of 28.8 kbps. This rate, while theo-retically attainable, is often not practical given the signal-to-noise ratio on unconditioned voice-grade telephone circuits.

The next rung up is Dataphone Digital Service (DDS), also called Digital Data Service. You connect to DDS using a special box called a Data Service Unit/Channel Service Unit (DSU/CSU). The DSU/CSU replaces the functions of the modem in the analog scenario. DDS provides speeds ranging from 2400 to 56 kbps. DDS lines are full-time leased connections between two specified locations, and sup-port a fixed bandwidth. They're usually used to construct private digital networks.

After DDS comes the Switched-56 (SW56) service, which lets you make dial-up digital connections to any other SW56 subscriber anywhere in the country (and often across international borders). It uses the same DSU/CSU as leased-line DDS, but includes a dial-ing pad for entering the telephone number of the remote SW56 system.

At the top end of the ladder are the truly high-speed digital services: Fractional T1 (FT1), T1, and T3. These services have an interesting history (discussed later), but for now you only need to know that the supported speeds are 64 kbps to 1.544 Mbps for FT1, 1.544 Mbps for T1, and 44.736 Mbps for T3. You connect to these services using a T1 or T3 multiplexer.

Switched & leased analog phone lines

Switched and leased phone lines carry digital data by first converting it to an analog signal using a device called a modem (from **mod**ulator–**dem**odulator). Modems have been with us for a long time, progressing continuously in speed and reliability. But modems have now reached a point beyond which further speed improvements are virtually impossible.

While computer processor speeds quadruple every five years, telecommunications speeds make similar jumps only once every decade. In 1963, the state of the art was 300 bits per second (bps). The year 1973 saw a blistering improvement to 1200 bps, jumping to 4800 bps in 1983 and finally just topping 14,400 bps in 1993. The latest modem technology reaches speeds of 28,800 bps, but that speed is obtained only under the best conditions.

Today's modems come in a bewildering mix of speeds and capabilities, which results in many compatibility and reliability problems. These problems, along with inherently low data rates, all but rule out analog links as a practical LAN interconnection. However, you're likely to see modems advertised with data rates at least as high as those of many purely digital services. Does this mean that modems can perform just as well as digital services? Well, yes, they can under certain conditions. With a leased line and high-speed modems from the same manufacturer, you can get effective throughput close to that of low-end digital services—about 56 kbps. Unfortunately, modem makers are claiming speeds double this figure, and leaving the impression that such speeds are routinely achievable. They aren't. Once you understand the truth about such claims, you'll see why modems probably shouldn't be on your list of internetworking alternatives.

Modem speed

Modem speeds claimed by manufacturers tend to be exaggerated all out of proportion to reality. Manufacturers often advertise the highest apparent speed of their products, regardless of how fast they actually pump data. When reading such vendor claims, keep in mind the following fact: for dial-up modems, the fastest practical speed today is 28,800 bps—a limit that is a function of the public switched telephone network's (PSTN) signal-to-noise (S/N) ratio. Shannon's Law is the information theory precept describing how much digital information can pass through an analog channel of a certain bandwidth with a given signal-to-noise ratio. The formula for that law is:

$$C = BW \log_2 (1 + S/N)$$

where C is the digital channel capacity in bits per second, BW is the bandwidth in hertz, and S/N is the signal-to-noise ratio. Most of the U.S. PSTN has a reliable signal bandwidth of 3 kHz and an S/N ratio of about 1000:1, yielding a maximum data rate of about 30,000 bps. Today's so-called v.34 or V.fast modems, operating at 28.8 kbps, have effectively reached that limit. How, then, does one explain the 38,400, 57,600, and even 115,200 bps claims made by modem makers? There is an easy answer to this question: data compression.

CCITT recommendation V.42 defines an error detection and correction protocol for modems that lets the modems themselves ensure reliable, error-free data transport. A modem with V.42 capabilities is a handy thing, because it removes from the attached computers the responsibility of routine error handling. Given that such modems provide guaranteed data transport, CCITT decided that the modem was also a good place to perform data compression, and released the V.42bis (from the Latin *bis* for *second*) recommendation in 1990. The data compression algorithm used in V.42bis modems has the potential to achieve as much as a four-fold reduction in data volume. In real life, though, compression depends on the data; only in rare cases does it ever reach even 50%, and then only with plain text data. The more forthright vendors report their 28,800 bps modems as running at 28.8 kbps, because that is their actual signaling rate. In typical networking applications you can expect to achieve modest amounts of data compression, which increases the effective throughput from 28.8 kbps to somewhere near the DDS and SW56 maximum speeds of 56 kbps.

There is another reason to look upon modem compression with a jaundiced eye: it turns out that the modem wasn't such a good place to put compression after all. It seemed like a good idea back when computer users ignored security and plain text was most often the data of choice. Now, however, users are turning to host-based encryption and compression both to protect their data and to get better data reduction. Host-based compression algorithms have advanced beyond the original V.42bis recommendation, and generally give higher compression ratios than modems. Also, to save space, users want to store files in on-line archives already in compressed form. Thus, most file transfers today are encrypted or compressed (or both) by the host, and cannot be compressed any further by the modem. In fact, modem compression actually increases the amount of data when sending a previously compressed file! Thus if your network activity consists mostly of file transfers where the files are already compressed or encrypted, you won't see any advantage from V.42bis.

Use as a backup medium

One practical use for modems in modern MAN/WAN networks is for network-management backup. While modem data rates are too low to carry much LAN data, they are adequate for performing network management chores, such as device control, performance measurement, and configuration maintenance. Network management traffic (e.g., Simple Network Management Protocol, or SNMP) usually rides on the same WAN facilities as LAN data traffic. If a WAN link fails, network management commands are cut off along with the data. Many WAN devices—bridges, routers, multiplexers, and concentrators—have an option for switched-network backup. When a WAN link dies, the device automatically falls back to a dial-up modem to maintain some network functionality. This lets you continue managing the WAN, perhaps reconfiguring it to use alternative wideband facilities.

Digital Data Service

Digital signaling offers much more bandwidth than analog signaling, and at higher reliability. By eliminating the conversion of digital data to an audio signal and back again, a digital signaling system eliminates many of the problems modems must deal with: audio noise, phase and frequency shift, clock synchronization, variable line quality, and signal attenuation. The electronics for attaching DTE devices to a digital link are also much less complex, which in the end means much less expense for equivalent bandwidth.

The most basic of digital services is called *Dataphone Digital Service* (DDS). The original name came from Bell Telephone's pre-deregulation days, when to use DDS you had to lease a special phone, called a DataPhone, from Ma Bell. With divestiture, these days are happily behind us; because DDS works with a wide variety of third-party equipment, the acronym DDS now stands for *Digital Data Service*.

DDS links are leased, permanent connections, running at fixed rates of 2400 bps, 4800 bps, 9600 bps, 19.2 kbps, or 56 kbps. A DSU/CSU device at each end provides the interface between the two-wire DDS line and traditional computer interfaces such as RS-232. A typical LAN interconnection uses two DDS-compatible bridges and external DSU/CSUs (Fig. 8-2).

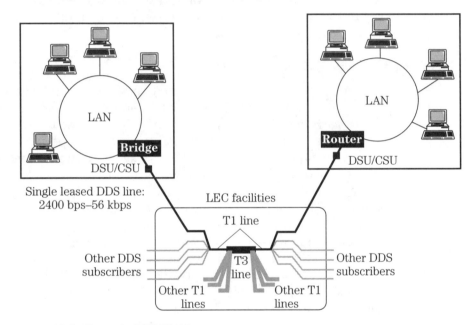

■ **8-2** *Example DDS WAN*

Inside the CO, a DDS line is merged into the regular flow of traffic on T1 and T3 carrier facilities, which route it to its destination. The DDS route is established when you buy the service, and bandwidth on the necessary trunk carriers is carved out at that time. You pay a fixed monthly fee for DDS, plus mileage charges based on the interoffice distance (as the crow flies) traversed over telco trunks. The requested data rate determines the fixed monthly fee.

DDS has physical limitations primarily related to the distance between the CSU/DSU and the serving CO. DDS works reliably when the route distance between the subscriber and the CO is less than 30,000 feet. Route distance is not mileage as the crow flies, but rather distance as the cable lies. An office seemingly only a mile or two from a CO may nevertheless have 20,000 feet or more of intervening cable, due to the circuitous routes local loops often take in metropolitan areas. Most telcos use a designed line when providing DDS service: telco engineers trace the shortest possible route over existing copper facilities to get from the CO to the subscriber's location. Telco field technicians then visit cable vaults along the route to make the necessary physical connections establishing the designed route.

Telcos accommodate longer distances in two ways: by adding signal extenders to the local loop, or by using two local loops (the so-called four-wire circuit) to improve signal quality and reduce noise. You'll likely pay extra installation and monthly charges when either of these measures is necessary. And despite even these heroic efforts, DDS will sometimes be simply inadequate for the distances required; for example, in rural areas distances run to tens of miles rather than thousands of feet—too far for DDS signal enhancers.

If you're internetworking two sites served by the same CO, a DDS line may be a practical solution. You get a continuous private connection at up to 56 kbps for about a hundred dollars per month. If your MAN spans two different COs, however, mileage charges can quickly add up to make DDS impractical. In such situations, you're better off with Switched-56 service, described next.

Switched-56

When you don't need a full-time connection, you can save money by using switched digital service, generally called Switched-56 (SW56). As its name implies, SW56 is only available at the fixed rate of 56 kbps. An SW56 link is similar to a DDS setup (Fig. 8-3): the DTE connects to the digital service via a DSU/CSU. An SW56 CSU/DSU, however, includes a dialing pad by means of which you can enter the telephone number of the destination SW56 station.

SW56 uses the same phone numbers as your local telephone system, and usage charges are the same as those for business voice calls (typically metered at .01 per minute). You can use SW56 for long-distance links, because your SW56 call is carried over the long distance digital network just like a digitized voice call. You can choose the long distance company to service your SW56 calls just as you do for voice calls. With the advent of competing switched digital services, such as Frame Relay Service, the cost of SW56 is falling—to less than $100/month in most areas.

You might be wondering at this point why you can't also get a 64-kbps switched digital service. After all, the underlying digital net-

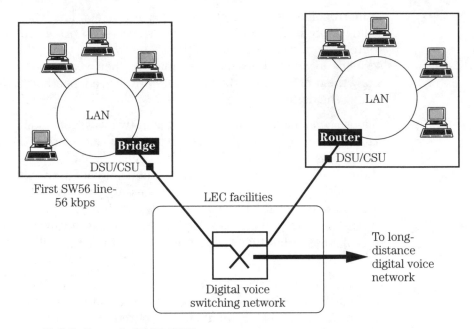

■ 8-3 *Example SW56 WAN*

work uses the 64-kbps DS0 channel to carry an SW56 call, so shouldn't that accommodate a 64-kbps data service? The answer is no, because a DS0 must carry call control information in addition to the data itself. Only seven of every eight bits transmitted carry data—the remaining bit carries the line state (off- or on-hook) and dialing information. This kind of signaling technique is called *in-band signaling*, because the call management information is carried in the same channel as the payload. Obviously, in-band signaling results in significant efficiency losses. ISDN (discussed in Chapter 9) overcomes this problem by using *out-of-band signaling*—a separate data channel for call management.

SW56 has applications in video and voice, too. Many teleconferencing codecs (from **co**mpressor-**dec**ompressor) combine two or more SW56 lines to obtain 112-kbps, to 384-kbps bandwidth for sound or video. Other codecs are available for transmitting high-fidelity audio (384 kbps is sufficient for CD-quality sound transmission), and voice teleconferencing systems.

SW56 lines are cheap enough that some networks use them to provide a primitive bandwidth-on-demand capability for long distance WANs. Figure 8-4 illustrates a network in which two offices, one in Chicago and one in New York, need to bridge their two LANs. Most of the time, interoffice LAN traffic is light, and a single SW56 con-

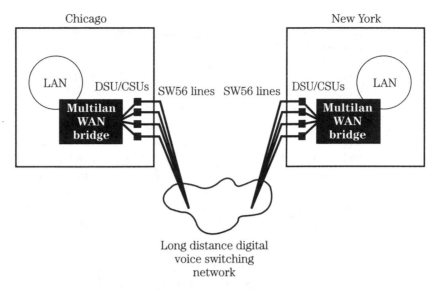

Chicago   New York

LAN   DSU/CSUs SW56 lines   SW56 lines DSU/CSUs   LAN

Multilan   Multilan
WAN   WAN
bridge   bridge

Long distance digital
voice switching
network

■ **8-4** *Example bandwidth-on-demand SW56 WAN*

nection is adequate. There are peak traffic periods, though, where bandwidth of up to 200 kbps is required. A LAN bridge feature called dynamic bandwidth adjustment solves this problem. Each bridge connects to multiple digital lines—in this case four SW56 lines. When traffic conditions warrant, one of the LANs dials additional digital calls to the other bridge, combining the bandwidth of all calls in 56 kbps increments as needed. With four SW56 lines, bandwidths of 56 kbps, 112 kbps, 168 kbps, and 224 kbps are possible. The calling bridge is configured to keep the call up for a certain length of time after traffic volume drops, to avoid repeated calls when the average traffic volume is high but not continuous. Bandwidth-on-demand bridges and routers frequently have the CSU/DSU devices integrated, obviating the need to buy and interconnect several boxes.

T1, Fractional T1, & T3 services

As noted earlier, telcos originally kept their digital trunks to themselves, to carry digitized voice traffic. Today, however, you can lease T1 and T3 trunks, obtaining pure digital pipes having bandwidths of 1.544 Mbps or 44.736 Mbps, respectively. Because of the reduced equipment and maintenance costs associated with T1 and T3 lines, telcos can provide them more cheaply than they could provide a 24-pair analog trunk. At the CO end, such an analog trunk must terminate at 24 separate line-service units to provide individual call handling. With a T1 line, all call management functions are provided by the customer's own equipment, either a digital PBX or a T1 multiplexer.

Many digital PBXs take advantage of T1 service to import multiple call circuits inexpensively. Instead of having 24 wire pairs terminated at a punchdown block and then routed into the PBX switch, a single wire pair carrying a T1 signal feeds the PBX, which then allocates as many 64-kbps DS0 subchannels for voice as necessary. In this scenario, the PBX is performing the voice digitization chore normally handled by the CO. The DS0 subchannels can be routed to other stations within the same switch, to stations at another PBX switch in a branch office, or to the public switched telephone network (PSTN). The PBX can use individual DS0 channels for data too, combining them as necessary for greater bandwidth.

As with a DDS circuit, you lease a T1 circuit between two locations. Unlike a DDS circuit, though, you have the ability to partition the bandwidth into multiple 64-kbps channels, taking the responsibility for establishing calls and performing other traffic management chores away from the CO. Figure 8-5 illustrates this by showing a typical four-office phone network built using (a) individual DS0 lines and (b) T1 lines. The DS0-line design requires preallocating a number of lines—called TIE lines—between offices; these carry "inside," interoffice calls. Another group of DS0 lines must be allotted to each office to access the PSTN for "outside" calls. As traffic patterns change, lines must be moved or added between offices, a time-consuming and inconvenient task.

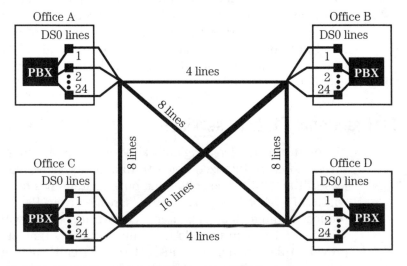

■ **8-5a** *Four-office phone network using DS0 lines*

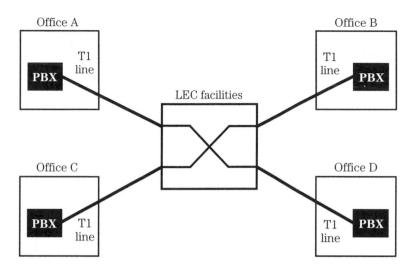

■ **8-5b** *Four-office phone network using T1 lines*

The T1 design consolidates interoffice communications into a single backbone consisting of four T1 lines, which is capable of meeting changing traffic demands on the fly. Because T1 service includes the ability to switch DS0s within the local network (the local CO switch), any DS0 in one office can be switched to any DS0 in another office or to the PSTN.

Data use of a T1 takes advantage of these abilities via the T1 multiplexer, a network terminating device that plays a role similar to that of an SW56 DSU/CSU. The difference is that the T1 multiplexer handles 24 DS0 channels instead of just one.

A T1 multiplexer splices up to 24 64-kbps DS0 channels into the single fat T1 1.544 Mbps bit pipe. The multiplexer uses time division multiplexing (TDM) to accomplish this, using a data framing format called *D4-Extended frame and format*, as shown in Fig. 8-6. Each frame contains 24 eight-bit bytes, with one byte for each DS0 channel. That works out to 192 bits. An extra trailing bit at the end of the frame carries out-of-band signaling information to pass dialing information and line status on to the switch. The frame is thus 193 bits long, formatted as 24 bytes plus one framing bit (this is where the expression "frame and format" come from). Each byte in the frame represents a time slot for the channel, with 8000 times slots per channel per second, and the product of 8000 and 193 yields 1,544,000 bits per second, which is where T1's 1.544 Mbps data rate comes from.

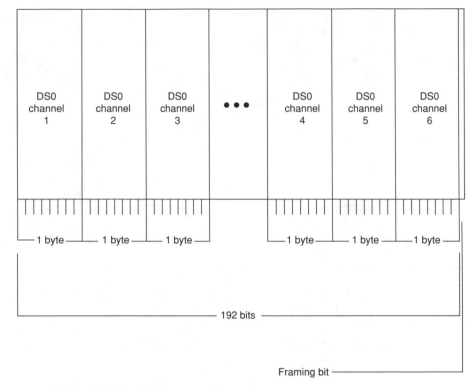

■ **8-6** *D4-Extended frame and format*

The T1 multiplexer simply moves bytes from the D4 frame into the appropriate DS0 channel, and vice versa, to accomplish the DS1/DS0 mapping. While this task must be performed very quickly (each frame must be processed in less than 125 microseconds), it isn't a complex operation. Hence T1 multiplexers are not terribly sophisticated devices. The T3 multiplexing process is identical, with the exception that frames are larger (i.e., 28 T1 frames) and slightly more complex.

Why should you need to know this? Well, beyond giving you a better appreciation for how telcos deliver T1 and T3 services, you need to know that this is the way T1 worked after AT&T fixed it. Perhaps you wondered what the "Extended" in "D4 Extended frame and format" meant? It refers to the revised T1 specification that made possible the use of the full DS0 channel bandwidth for data. The previous D4 nonextended (busted) frame and format standard supported only 56 kbps worth of data per DS0, or seven bits per byte. The eighth bit—more than 12% of the total bandwidth—was reserved for call supervision. This was a terrible waste, because that signaling bandwidth was actually only needed at the beginning and end of a call; while a call was in progress, the eighth bit went empty. Worse, though, the old D4 format transmitted dialing information by stealing bits right out of the middle of the data stream, a technique appropriately called robbed bit signaling. In a voice call, a bit lost here or there is never noticed, but in a data call such arbitrary corruption causes data errors that must be resolved by higher-layer protocols. The D4-extended format permits all eight bits of a byte to carry data, and multiplexes dialing and call supervision signals into the 193rd bit of each frame. That 193rd bit moving down the T1 8000 times per second provides an 8 kbps signaling channel, which is easily fast enough for call-management traffic.

T1 and T3 circuits are useful MAN tools because they offer adaptable bandwidth at essentially fixed costs within a metropolitan area. T1- and T3-capable routers and bridges typically support one or more T1 or T3 circuits, and automatically make connections with other routers or bridges in the network. Figure 8-7 illustrates a typical T1-based MAN. As a network designer, you program the router or bridge to specify the adjacent routers or bridges for each digital circuit. You can also specify bandwidth-on-demand parameters to change bandwidth in DS0 increments, similar to the technique used with SW56 lines. This lets you share DS0 increments with a PBX or video-conferencing codec. Figure 8-8 shows such a network.

Fractional T1

Leasing a T1 line means paying for the entire 1.544-Mbps bandwidth 24 hours a day, whether it's used or not. A relatively new service, called *Fractional T1* (FT1), lets you lease any 64-kbps submultiple of a T1 line. You might, for example, lease only DS0 through DS5 to obtain six 64-kbps channels or an aggregate bandwidth of 384 kbps. FT1 is useful whenever the cost of a dedicated T1 would be prohibitive. It's not as efficient or as flexible as fast-packet services, however, because you're paying to have the fraction of bandwidth you've leased available on a 24-hour basis. However, FT1 has an intrinsic feature not available with full T1 circuits: multiplexing DS0 channels outside your own enterprise T1 network.

Because you're not leasing an entire T1 circuit, you can't dictate the location of the other end of the circuit. After all, you're going to be sharing the T1 with other customers. The remote end of an

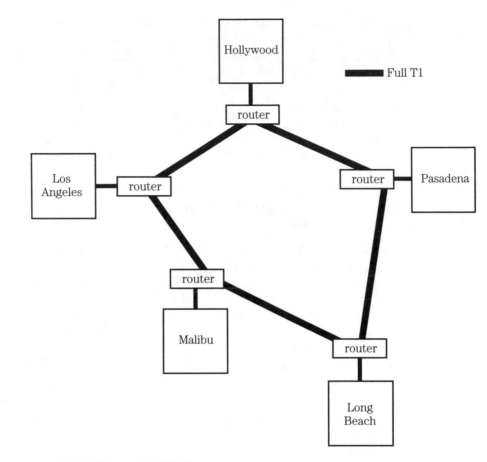

■ **8-7** *Example T1 WAN*

FT1 circuit is at a telco-managed *Digital Access Cross-Connect Switch* (DACS). Everyone leasing an FT1 has the far end embedded in the DACS, where the telco has established its own network of T1 interconnects. Any two companies sharing the same DACS can switch among each other's DS0 channels, provided that the telco has configured the two companies as interoperating organizations. This interoperability can be an advantage when a large central organization (e.g., a government agency) needs to interoperate with many smaller organizations (e.g., contractors).

Due to the one-ended nature of FT1, you must lease a separate FT1 circuit for each node in your network. Contrast this with T1 circuits: you need only lease one between each pair of nodes. For this reason, as the size of the fraction increases, FT1 eventually gets more expensive than a T1. Typically, the threshold for this cost reversal has been around 75% (i.e., eighteen FT1 circuits).

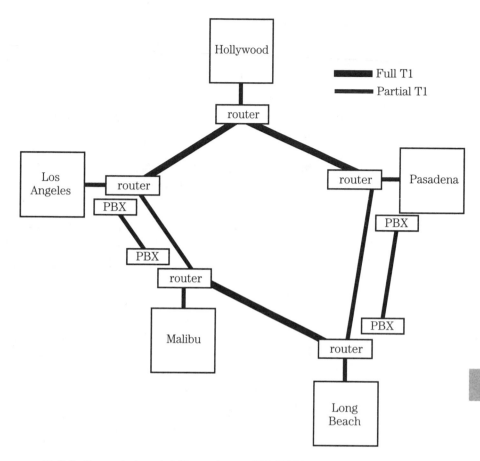

■ 8-8 *Example bandwidth-on-demand T1 WAN*

Merging into FastPacket

As T1 became more prevalent, telcos began honing methods of delivering service more cheaply (for themselves) and efficiently. FT1 is an example of this honing. Another example is T1 FastPacket (T1-fp) service.

The traditional T1-TDM technique results in a lot of wasted bandwidth. In voice calls alone, a significant amount of transmission time is wasted sending silence—brief spaces between words and sentences. LAN traffic, too, has periods of inactivity. T1-fp uses packet switching technology instead of TDM to share the T1 bandwidth, allocating bandwidth on an as-demanded basis to a set of virtual circuits that replace the dedicated circuits of T1-TDM.

By suppressing silence in voice conversations and idle characters in LAN traffic, T1-fp makes considerable improvements in line utilization. For example, a typical voice call contains 60% silence; stripping out the silence in two voice calls lets them both be carried simultaneously over a single 64-kbps fraction. LAN traffic can show even higher efficiency improvements.

Figure 8-9 compares T1-TDM and T1-fp. T1-TDM allocates a separate slot in each frame to each of the 24 discrete DS0 channels, while T1-fp dedicates the entire frame to one user, tacking a destination address at the end. With the whole frame dedicated to a single user, the frame need not be disassembled by the switch; instead, it's routed *en masse* to its single destination.

T1-TDM

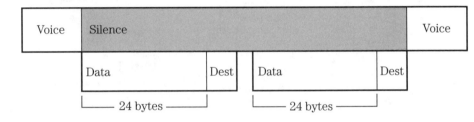

Byte

T1-FastPacket

■ **8-9** *T1-TDM and T1-fp compared*

Here we come full circle to join the technologies discussed in Chapter 7. T1-fp was an innovative development and important conceptual leap. It led directly to the invention of Frame Relay Service and eventually to ATM and Cell Relay Service. As these more advanced services become widely available, T1-fp will fade into the background. In the meantime, it serves as a useful transitional tool for networks that outgrow T1 capacities but aren't yet ready for T3.

Multipoint links

9

Introduction to multipoint links

As network interconnections become more common, companies are discovering the need to internetwork with outside organizations. Some outside interconnects are short-lived and infrequent, as in temporarily attaching a remote workstation to provide some sort of data delivery service (e.g., database search and retrieval). Other interconnects are more persistent, such as joining the LAN of a manufacturer with closely affiliated suppliers to conduct frequent Electronic Data Interchange (EDI) dialogues. One of the fastest-growing outside-organization interconnects is that of a corporate LAN to the worldwide Internet through an Internet service provider.

The ad hoc nature of these sorts of interconnections makes dedicated point-to-point links impractical. By the time such links can be installed, the need to interconnect has often passed. Certainly wideband links such as T1 and T3 circuits require a great deal of justification—in the form of frequent important traffic—to set up and maintain. But even the more flexible SW56 service is impractical for many spur of the moment network marriages, because each SW56 circuit requires a dedicated telephone number, leaving the users of SW56 service to arbitrate amongst themselves for available channels. This is a little like a business having a multiline phone system with no automatic "hunt" feature; every incoming call would have to be to a different phone number!

The obvious solution to this problem is to provide digital services similar to those of a multiline PBX, with multiple incoming and outgoing calls permitted to and from a single location. Such services are available; they use multipoint links to establish arbitrary connections between organizations that may not have a deeply intimate relationship.

History of multipoint links

The oldest and most prevalent multipoint link is the *X.25 Public Switched Data Network* (X.25 PSDN). Using X.25, you can establish a connection virtually anywhere in the world. As an internationally accepted protocol, X.25 is the single commercial networking technology that operates across foreign borders. None of the traditional point-to-point links—DDS, SW56, T1, FT1, T3, or SMDS—can offer international connections today. That's largely because these services have (sometimes wildly) conflicting standards that rarely interoperate. For example, T1's data rate of 1.544 Mbps clashes with the European equivalent service called E1, which operates at 2.048 Mbps. X.25 does have its flaws: with a maximum speed of 56 kbps, it's slower than a slow boat to China. However, it's better than nothing, if nothing is your only alternative.

The next most common multipoint link is *Integrated Services Digital Network* (ISDN). ISDN falls under the more comprehensive *Broadband Integrated Services Digital Network* (B-ISDN) standard, which also covers ATM, SONET, and a few other soon-to-be worldwide standards. ISDN availability is relatively new: ready access is available in most major metropolitan areas, but interoperability between metropolitan areas has been supported only recently. U.S. ISDN (also said to mean "It Still Doesn't Network") does not yet interoperate with European ISDN, although the standards say that it should eventually. Basic ISDN service provides two 64-kbps data channels and one 16-kbps signaling channel that can also, incidentally, carry X.25 traffic.

The newest (and least available) multipoint link is *Switched Multi-megabit Data Service* (SMDS). SMDS is a metropolitan area network (MAN) technology designed by Bell Communications Research (Bellcore), the R&D arm of the Regional Bell Operating Companies (RBOCs). Still in the early stages of deployment, SMDS initially will provide access only between systems within a metro-sized geographical area. The standard allows, however, for connections between any systems that have an SMDS address, anywhere in the world. SMDS supports data rates of 1.544 Mbps (DS1) and 44.736 Mbps (DS3), often rounded up to 45 Mbps for simplicity), with the vast majority of users connecting at the DS1 speed.

X.25 packet switching networks

Packet switching networks accept data streams from many different users, divide each data stream into discrete packets, and then route the packets to destinations over a shared transmission network. X.25 is the mother of all packet switching technologies, as evidenced by its wide availability and mature set of standards. X.25 interfaces are cheap and plentiful, and X.25 PSDNs span the globe. X.25 has its limitations—a maximum data rate of 56 kbps and variable response time due to packet delays—but its ubiquity makes it a useful WAN tool when other technologies aren't available.

Packet switching basics

An X.25 packet-switched data service accepts packets from an end user application, transports them through a network of computers, and delivers them to a specific end user application. Packet switching's advantages over point-to-point links are the effective sharing of local transport and wideband carrier facilities, and shorter connection setup times. The disadvantage is the variable delay that packets experience en route to their destination.

X.25 is both a networking standard and a set of communication protocols. The networking standard defines the devices and paths that connect to form an X.25 network—local access components, packet nodes, network links, and network management systems. The protocols define how data moves across the network of devices, with separate protocols for local access, internode traffic, and network management. Figure 9-1 illustrates the components of an X.25 network. The components and their functions are the following:

☐ Local Access Component (LAC). LACs provide the interface between computer or terminal equipment—collectively called Data Terminal Equipment (DTE)—and the X.25 network. DTEs connecting to a LAC do not know anything about X.25 protocols—the connection appears as a pure bit pipe with some arrangement for establishing connections with a remote endpoint. A LAC is often an RS-232 link to a PAD (described below). It may also be a dial-up or dedicated connection over a modem to a packet-capable LAC. The LAC can obtain connection information from the DTE in any way it deems necessary—the LAC interface is not standardized. Typically the DTE sends an ASCII command string to the LAC to establish a connection. A LAC is only necessary when attaching non-X.25 devices to the network.
☐ Packet Assembler/Disassembler (PAD). A PAD is a small computer that converts non-X.25 data streams into X.25 packets, and vice versa. The PAD provides access to the

network for LACs or X.25-savvy devices. The PAD also performs call setup, protocol conversion and emulation, and rate adaptation (adapting variable DTE data rates to the fixed X.25 access rate). A PAD can be a standalone device with a built-in LAC for DTE equipment and a digital data line connection to the X.25 network, or it can be integrated into a personal computer or host system in the form of an X.25 adapter card or software driver. In the latter case, the PC itself is the computer performing PAD functions. A standalone PAD can be collocated with the DTE equipment, or might be remotely located, with DTE access via switched or leased telephone lines and modems. An example of the latter is the CompuServe dial-up X.25 network.

☐ Packet Node (PN). Packet nodes make up the bulk of an X.25 network; they accept packets from PADs and route them to the proper destination PAD. PNs also perform performance analysis and traffic balancing, collect billing information, and ensure reliable transport by retransmitting bad packets. Some PNs provide gateways to other networks, such as the Internet. Each PN consists of a computer running packet-switching software, various network interfaces, and possibly some disk storage for archiving copies of software, as well as statistical and billing information. The connection between two PNs is called a Network Link (NL).

☐ Network Link (NL). Every PN has at least one circuit linking it to another PN; that logical circuit is the network link. The circuit may be a physical connection, such as an analog or digital leased line, or a logical connection, such as a channel over a microwave, satellite, or fiber link. NLs are permanent connections, ranging in speed from 9600 bps to DS1 (1.544 Mbps). The network of PNs and NLs make up the fabric of an X.25 network.

☐ Network Management System (NMS). The NMS oversees the population of PNs and PADs. It maintains the database of nodes and interconnects, a table of possible routes, and profiles describing all authorized end users. An NMS also contains copies of all PN software, and has the ability to download upgraded software directly to a PN. That capability, combined with the ability to remotely monitor the traffic flowing through every PN, lets an X.25 network be managed without on-site personnel at each PN location.

X.25 supports network interface speeds of 2400 bps to 56 kbps. Unlike more modern fastpacket technologies, X.25 also guarantees re-

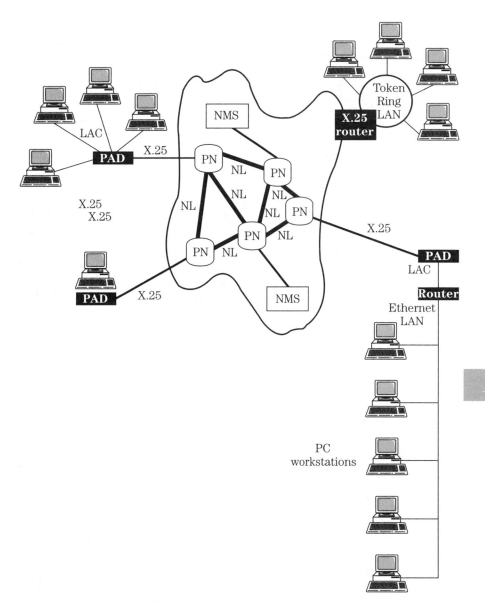

■ 9-1 *WAN using dedicated digital circuits*

liable transport. This means that every packet is acknowledged at every waypoint along its entire route, and any errors are corrected by the network retransmitting the packet (Fig. 9-2). Each station along the route stores a packet until it gets a positive acknowledgment from the adjacent station. This imposes considerable protocol overhead for error control information. X.25's error detection and correction process is necessary, given the low reliability of the

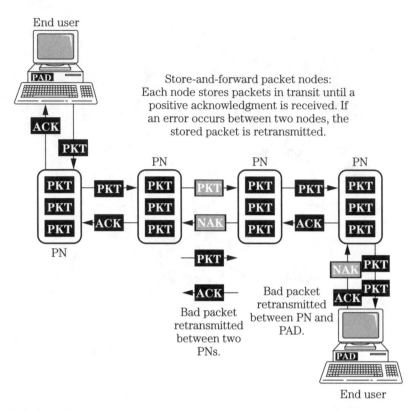

■ **9-2** *X.25 guaranteed packet delivery*

copper and microwave long distance networks it originally used. Today, X.25 rides on the same all-digital long distance facilities (T1 and T3) used for digital voice transmission. The inherent reliability of the current digital carrier network eliminates the need for X.25's error correction. X.25's replacement technologies—frame and cell relay—don't perform error detection or correction at all, leaving that task to higher protocol layers. Higher-protocol error recovery is slower than recovery in the data link itself, but because errors are infrequent the increased delay is insignificant. In the meantime, packet routing is faster, because nodes need not store packets for possible retransmission, and decreased protocol overhead reduces the amount of unproductive data traffic.

Virtual circuits

X.25 uses the concept of a virtual circuit (VC) to provide logical data paths between endpoints in the network. The idea is straightforward: each packet has a header associated with it that contains a *virtual circuit identifier* (VCI), which associates the packet with an established logical connection, or VC (Fig. 9-3). The header is added to the packet when the packet is created by the PAD. Many VCs can exist over a single physical connection, up to the maximum numerical value of the VC field in the packet header. X.25 supports up to 16 groups of 256 virtual circuits each, for a total possible VC count of 4096. In X.25, a VC is called a *Logical Channel* (LC); each group of LCs is called a *Logical Channel Group* (LCG). The number assigned to each LC is called a *Logical Channel Number* (LCN). LCs can be permanent (PVC), with the virtual connection (and route through the packet switch) established once by the service provider, or switched (SVC), established via a call setup procedure by the end user every time the

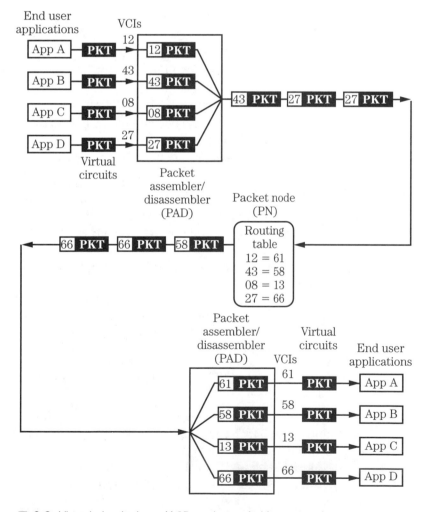

■ **9-3** *Virtual circuits in an X.25 packet switching network*

circuit is used. PVCs are like leased lines, and have fixed LCN assigned at installation time. SVCs get a new unique LCN whenever a new connection is requested; they're similar in concept to switched dial-up lines. Note that LCNs have local significance only—the LCN on the other end of an LC will likely have a different value. This isn't a problem, though, as LCNs simply keep LCs separate at the network interface. The internal network will use different LCNs for each hop in the route.

When a PN receives a packet, it looks up the local LCN in a routing table to determine the LCN of either the destination or the next hop in the route. The LCN in the packet is then changed to the new value and the packet is transmitted to the next stop on the route.

X.25 packet format

An X.25 packet consists of a fixed-length header followed by a variable number of optional fields, the payload bytes, and finally a checksum value. Figure 9-4 shows the packet layout for one kind of X.25 packet, the data packet. Other X.25 packet formats have additional header fields; these packet types are used to initiate and terminate connections, and to route data internally on the network. The following header fields are used in every X.25 packet:

- ☐ General Format Identifier (GFI). Indicates the format of the X.25 packet. For data packets, the value is 0001 (binary).
- ☐ Logical Channel Group (LCG). Identifies the logical channel group for this packet. Valid values are 0 through 15 (decimal).
- ☐ Logical Channel Number (LCN). Identifies the logical channel number for this packet. Valid values are 0 through 255 (decimal).
- ☐ Packet Type Identifier (PTI). Indicates the type of X.25 packet (e.g., request or indication, call setup or termination, data or control). The PTI determines the format of the variable part of the packet.

Data packets have additional header fields related to packet sequencing and error control:

- ☐ P(R): Packets Received counter. A wrapping counter with values from 0 to 7 (decimal) indicating the sequence number of the last packet received.
- ☐ M: More data flag. When 1, indicates that more packets follow, and must be treated as a unit by higher protocol layers.

204

Fixed portion of header   Fields for data packets

GFI	LCG	Logical channel number (LCN)	Packet type identifier (PTI)	P(R) (3 bits)	M	P(S) (3 bits)	0	Payload	Frame check sequence (16 bits)
(4 bits)	(4 bits)	(8 bits)	(8 bits)					(1 to 128 bytes)	

— 1 byte — 1 byte — 1 byte — 1 byte — 2 bytes —

■ **9-4** *X.25 data packet format*

☐ P(S): Packets Sent counter. A wrapping counter with values from 0 to 7 (decimal) indicating the sequence number of the last packet sent.

☐ Frame Check Sequence. A longitudinal checksum of the entire packet, used to detect errors. Error correction is via retransmission of the entire packet.

The P(R) and P(S) values make up the X.25 mechanism for ensuring that packets are transmitted reliably and in sequence. These counters provide a windowed acknowledgment scheme in which up to eight packets can be in transit at any one time before the sender must wait for an acknowledgment. At each hop in the route, the receiving node verifies the checksum and, if the check fails, asks for the packet to be retransmitted by sending an REJ (reject) packet to the sending node.

Connection management

Connection management is handled by the PAD, which obtains from the DTE all the information necessary to establish a connection and performs packet assembly and disassembly once a connection is established.

In addition to data packets, X.25 has special packets for establishing and clearing a call. Figure 9-5 illustrates a typical conversation between two endpoints. Every connection operates in three phases: call setup, data exchange, and call termination. During the call setup phase, the requesting DTE sends a *Call Request* packet, identifying the address of the destination DTE. The DTE receives an *Incoming Call* packet identifying the caller and type of service requested. If the DTE accepts the call, it returns a *Call Accepted* packet, which the sender receives as a *Call Connected* packet.

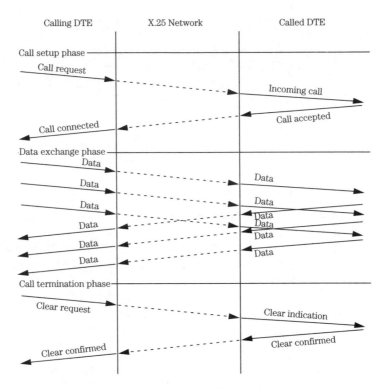

Calling DTE   X.25 Network   Called DTE

Call setup phase
  Call request
                                              Incoming call
                                           Call accepted
  Call connected

Data exchange phase
  Data
  Data   Data
  Data   Data
  Data   Data
                                          Data
  Data   Data
  Data   Data
  Data

Call termination phase
  Clear request
                                          Clear indication
                                          Clear confirmed
  Clear confirmed

■ **9-5** *X.25 call establishment and clearing sequence*

Once a call is connected, the parties can exchange any number of data packets in both directions. As noted earlier, the X.25 sequencing scheme provides a transit window that permits up to eight packets en route in each direction before the sender requires an acknowledgment.

Either side of the connection may decide to terminate the call by sending a *Clear Request* packet. The receiver of this packet should then send any enqueued packets, followed by a *Clear Confirmed* packet.

Interconnecting LANs with X.25

While X.25 is too slow for many LAN interconnect applications, it's better than nothing when no advanced networking technologies are available. X.25 connections are also valuable as a backup to higher-speed connections, ensuring that you have management access to network assets even when a wideband link is disabled.

Many WAN routers and bridges today support X.25 connections, both as primary and backup links. Some products can gang together several X.25 network connections to build up wider-bandwidth channels between LANs. In such situations, though, the actual throughput is not simply the sum of the ganged X.25 channel bandwidths. Packet delays, which force a router to wait for a group of packets across several X.25 channels, result in a measurable throughput that is somewhat less than the theoretical maximum.

Integrated Services Digital Network (ISDN)

ISDN promises to be one of the lowest-cost entry points for digital communications. It is positioned to completely replace modems in the existing public switched telephone network (PSTN), a feat it accomplishes by essentially moving the digitization of analog signals from the telco's central office (CO) to the customer's premises. ISDN is priced like regular analog phone lines, with each ISDN line delivering one 16-kbps and two 64-kbps channels. The typical ISDN connection costs about as much as two analog lines (i.e., $25 to $75/month). Usage is priced just like analog lines too: free local calls for residential users, and metered local calls (at rates of one to four cents per minute) for business users.

ISDN delivers a four-fold speed improvement over analog modems, and it accomplishes this using the existing copper local loop and premises telephone wiring. The "Integrated" part of ISDN's name refers to the combining of voice and data services over the same wires. This integration brings with it a host of new capabilities that combine voice, data, fax, and sophisticated switching. And because ISDN uses the existing local telephone wiring, it's equally available to home and business customers.

ISDN service is available today in most major metropolitan areas; it was more than 50% deployed by the end of 1994, and will probably be completely deployed throughout the U.S. by the end of 1995. The only physical constraint is that the customer must be within 18,000 feet of the central office switch, and even this restriction can be relaxed with the use of repeaters, which extend ISDN out to about 50,000 feet.

ISDN basics

To really understand ISDN's features, you'll need to know a bit of its terminology. Figure 9-6 shows a minimal ISDN setup connecting two computers. The figure also shows the external *ISDN reference points*, labeled R, S/T, and U. (Don't strain yourself trying to deduce what R, S, T, and U stand for—they are simply consec-

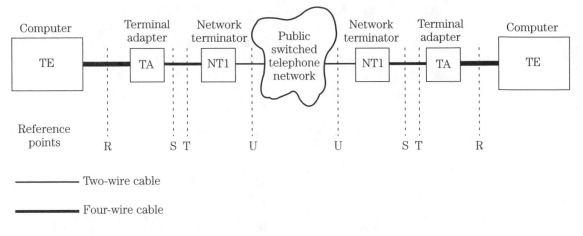

Computer | Terminal adapter | Network terminator | Public switched telephone network | Network terminator | Terminal adapter | Computer

Reference points
R   S T   U   U   S T   R

————————— Two-wire cable

————————— Four-wire cable

■■■■■■■■■ RS-232 cable

■ **9-6** *Simple ISDN connection between two computers*

utive letters of the alphabet, chosen by the ITU as the next available designations from the entire set of ITU standards). Each interface point requires an electrically different device connection and cabling.

The U reference point is where the incoming unshielded twisted pair enters a device called the *network terminator* (NT1), which breaks the 144-kbps two-wire channel into one 16-kbps and two 64-kbps subchannels over a four-wire balanced bus. The four-wire balanced bus provides better noise immunity than a two-wire carrier—necessary in the electrically noisy environs of building wiring.

The S/T reference point denotes this four-wire UTP cable. The two 64-kbps channels are called *bearer channels*, because they exclusively carry customer voice or data signals. The 16-kbps channel is a special data channel used for out-of-band signaling between your ISDN equipment and the phone company's central office, but it can also be used for X.25 network connections. The combination of two bearer channels and one data channel is called the *Basic Rate Interface* (BRI) in telco lingo, or sometimes just *2B+D* for short. You also can buy ISDN in bulk: 23 B channels, with a single 64-kbps D channel. This service, called the *Primary Rate Interface* (PRI), inherits most of the capabilities and limitations of BRI, so features of 2B+D apply to PRI's 23B+D service as well.

The S/T reference point actually denotes two separate reference points that are electrically identical (four-wire UTP). The distinction is that the T reference point denotes a point-to-point bus while the S reference point denotes a multipoint bus. The multipoint bus allows multiple devices to be attached to one ISDN BRI line.

The R reference point denotes the interface at which you connect non-ISDN equipment, such as analog phones and RS-232, so its characteristics are undefined.

Continuing with Fig. 9-6, the four-wire unshielded twisted pair cable carries the 2B+D channels to another device called the *Terminal Adapter* (TA). Unlike the NT1, which provides only a single function (materializing the 2B+D channels on the four-wire balanced bus), the TA can do many things. Its job is to connect various sorts of Terminal Equipment (TE)—computers, fax machines, or telephone sets—to one of the B channels. Depending on the variety of equipment you want to connect, the TA might be cheap or expensive, simple or complex. In this example the TA is shown as a separate unit, but it could easily be contained within the computer as an add-in card or integrated feature.

A typical TA for data-only applications might simply emulate a pair of ordinary (albeit very fast) Hayes-compatible modems, translating standard modem setup and dialing commands into ISDN call setup commands. You connect computers and other DTE (Data Terminal Equipment) devices to this kind of TA with a normal RS-232 cable, and use common modem or fax software set to 57.6 kbps. The TA provides automatic rate adaptation to match whatever data rate your computer supports with ISDN's 64-kbps channel, so that if the attached computer can't communicate faster than, say, 38.4 kbps, it will still work fine under ISDN (and even connect properly to a remote computer operating at some other speed).

Rate adaptation is not strictly part of the ISDN standard; instead, one existing and two new data link standards are superimposed on ISDN to accomplish speed matching. X.25 is an existing packet-oriented protocol adapted to ISDN via a special PAD (Packet Assembler/Disassembler) geared to generate packets within the constraints of the ISDN data stream. The two new rate adaptation standards are V.110 and V.120, which perform speed matching using bit stuffing and packet assembly/disassembly, respectively. V.110 can adapt DTE speeds of up to 19.2 kbps, and V.120 can adapt up to 64 kbps. All three rate adaptation standards—X.25, V.110, and V.120—are usually supported by modern ISDN TAs.

By replacing modems with a digital service, ISDN eliminates problems due to intermittent line noise, speed mismatches, and modulation conflicts. You can get more value out of ISDN by investing in advanced TAs or direct computer-integrated ISDN hardware. For example, while you're limited to two devices using the 64-kbps channels at any one time, up to eight devices can share access to the channels, because the four-wire network emerging from the NT1 acts as a passive bus. The passive bus uses a second kind of network terminator, called an NT2, to let up to eight separate TAs share a single 2B+D circuit. Figure 9-7 shows a passive bus with a dozen computers and four fax machines sharing an ISDN circuit. Each pair of TEs (computer or fax machine) requires one TA; whenever a TE wants to use a B channel, its associated TA checks to see if a channel is available, and if so, dedicates it to the requesting TE. The example shows maximum device sharing, but the cost of additional 2B+D circuits is low enough that you'll likely have fewer devices on a single bus.

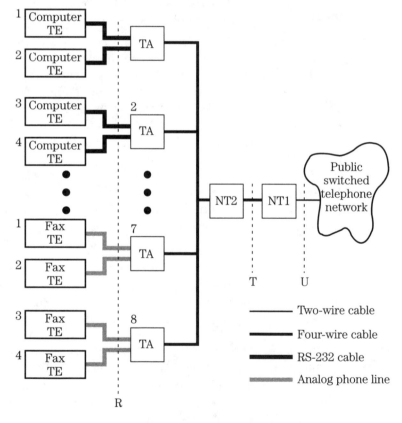

■ **9-7** *ISDN multipoint configuration using passive S/T bus*

210

You can avoid the use of external TAs altogether by connecting computers to ISDN directly, using plug-in ISDN adapter boards. Most such boards also incorporate an analog TA, letting you connect analog equipment such as modems, analog telephones, and fax machines. Along with the adapter, you'll need software supporting ISDN's call control protocols. Some ISDN boards also incorporate a digital modem, which emulates an analog modem by generating the same digital waveform that would result from an analog modem plugged into a TA analog port. By generating the digital waveform directly, these devices eliminate signal quality losses that occur in analog-to-digital conversion. Digital modems provide backward interoperability with non-ISDN-equipped computers and fax machines; because they use analog modem protocols, they're limited to the same speeds as analog modems.

The term "digital modem" is also used (erroneously) to refer to some all-in-one TA products designed as plug-for-plug replacements for modems. These devices have an integrated NT1, and thus connect directly to the ISDN U interface. The other connectors on such products are usually RS-232 ports emulating a Hayes modem. Because these devices can take the place of a modem in an existing application, they're frequently called "digital modems," apparently to avoid ISDN terminology.

ISDN uses the existing voice telephone network, and thus has all the attributes of the voice switching system. However, ISDN extends many of these attributes, because call control is performed out-of-band, on the D channel, rather than in-band using touchtones. Each ISDN line has a telephone number assigned, but that telephone number can support more than one call, or *appearance*, at a time (Fig. 9-8). A given ISDN line can support up to 15 call appearances, with any two of these connected to an active B channel at one time. Appearances that aren't connected are like calls on "hold"—they remain alive indefinitely until a D channel signal switches them to a B channel for active use.

When a call comes into an ISDN line, the central office switch sends a call setup message to the TA via the D channel (if multiple TAs are connected via passive bus, any TA can pick up the call). The call setup message contains a *bearer type* code indicating the type of incoming call, analog or digital. An appropriate TA can pick up the call and assign it to an available B channel, or if both B channels are in use, it can free a channel by placing an active call on hold and making the new call active. These calls can be either data or voice, in any combination. Thus, a single TA can have up to

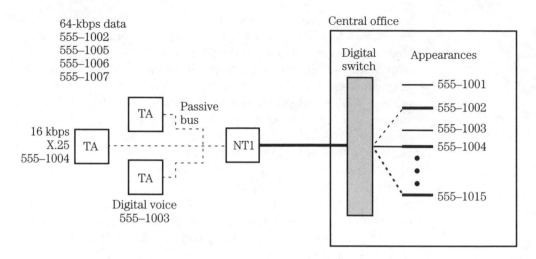

64-kbps data
555–1002
555–1005
555–1006
555–1007

Central office

Digital switch

Appearances

—— 555–1001

—— 555–1002

—— 555–1003

—— 555–1004

●
●
●

—— 555–1015

Passive bus

16 kbps X.25 555–1004

TA

TA

TA

NT1

Digital voice
555–1003

■ **9-8** *ISDN call appearances*

fifteen simultaneous calls in progress, with any two of those calls active (i.e., actually communicating).

As noted earlier, the ISDN D channel also supports X.25-based packet services such as Telenet and Tymnet. Such X.25 traffic co-exists with call control traffic between the TAs and the central office switch. Because X.25 supports multiple virtual circuits, you can open several simultaneous X.25 "calls" to different remote sites.

Connection management

Connection management is handled by the TE, which establishes a connection with the remote TA using a regular PSTN telephone number by sending the call setup data packet as described earlier. Each connection to a remote endpoint is thus a point-to-point connection. You might be wondering, then, why ISDN is considered a multipoint technology. The answer is that while each ISDN B channel can only communicate with one endpoint at a time, connections can be established and maintained with up to 15 different endpoints, or call appearances (Fig. 9-8). You can think of a call appearance as a sort of multiple call waiting feature. Each appearance can have a unique directory number (DN), or the entire BRI interface can have one DN and use an automatic hunt feature to attach incoming calls to a new appearance. On a given B channel, one appearance is active and the others are "on hold"—the connection is established and available to the TE at any time. A TE can switch between appearances at will.

212

ISDN's out-of-band signaling over the D channel makes this possible. When an incoming call arrives at the CO, the CO sends a packet down the D channel announcing the call's availability. Any TA on the passive bus can accept the call by putting an existing call on hold. Once the call is accepted, it enters the pool of current appearances within which the TA can switch.

ISDN's D channel signaling also enables a host of other features, such as call forwarding, caller ID, call routing, and automatic call rejection.

Either partner TA may decide to terminate the call by sending a *Release* packet over the D channel. Releasing a call simply discards that connection's call appearance—the B channel may or may not have been actively communicating using that appearance. In this way, call management is entirely asynchronous to data communications.

Interconnecting LANs with ISDN

At first glance, ISDN seems to suffer from the same speed limitations as X.25: its 64-kbps B channel data rate is not much faster than X.25's 56 kbps. But ISDN has some twists that mitigate this apparent bottleneck. First, ISDN lines are direct end-to-end bit pipes, with no packet routing and the concomitant packet transmission delays. Thus bits put in one end of an ISDN connection flow out of the other end at precisely the contracted data rate (64 kbps or, in some cases, 56 kbps). Second, multiple B channels can be ganged together using a technique called *bonding* to obtain any size aggregate channel up to the sum of 23 B channels in an ISDN PRI link. ISDN's timely bit delivery comes from the need to deliver audio data in real time; this same timeliness makes ISDN suitable for other digital services, such as video conferencing and CD-quality audio. Many video and sound codecs now support B channel bonding to acquire the bandwidth necessary for these services. In this way, ISDN is similar to fractional T1.

Even an ISDN BRI line can be bonded to form a 128-kbps channel, and many routers and bridges on the market have such bonding built in. An ISDN BRI line also makes an attractive backup connection for broadband links, because it provides reasonable throughput and the ability to achieve higher bandwidths on demand by combining multiple BRIs.

Switched Multimegabit Data Services (SMDS)

SMDS is a proprietary fastpacket service designed by Bellcore to provide a metropolitan area networking facility. In light of the rapidly growing deployment of Frame Relay Service (FRS), Cell Relay Service (CRS), and Asynchronous Transfer Mode (ATM) (discussed in Chapter 7), SMDS may in the future provide inexpensive MAN interconnect service. However, due to limited availability, SMDS will not likely be a general purpose near-term MAN solution. The ability of SMDS to provide data rates higher than the 1.544 Mbps of FRS make it a truly high speed public switched data service.

Initially, SMDS deployment will provide access only between systems within a metro-sized geographical area. The standard allows, however, for connections between any systems having an SMDS address, anywhere in the world. SMDS supports data rates of 1.544 Mbps (DS1) and 44.736 Mbps (DS3).

Although SMDS is a proprietary service, the underlying protocols and facilities conform to the IEEE 802.6 MAN standard. This standard uses the same packet format as CRS, and thus SMDS can interoperate over CRS.

SMDS basics

Figure 9-9 illustrates a typical SMDS network topology and its components. Because SMDS is a LAN interconnect technology, the primary end user device is a LAN router. The router connects to the SMDS network using an SMDS Access device, which uses a *distributed queue dual bus* (DQDB) high speed optical channel for transport to the LEC's SMDS switch; this transport is called the subscriber network interface (SNI). The SNI consists of two unidirectional fiber optic cables, called Bus A and Bus B, to carry asynchronous traffic between the SMDS switch and access unit.

The SMDS switch is a metropolitan switching system (MSI), designed to process SMDS traffic for a small metropolitan area—typically the same area served by a traditional telephone switch. Several MSIs in a region feed into a common Inter Exchange Switch (IXC). The IXC routes traffic from one MSI to another as required, performs the network management function, and manages the subscriber database. Eventually IXCs will be interconnected to collect multiple MANs into a WAN, providing worldwide access. That interconnection is not part of the SMDS standard, but because SMDS uses the same packet format as CRS, SMDS will interoperate over wide-area ATM networks.

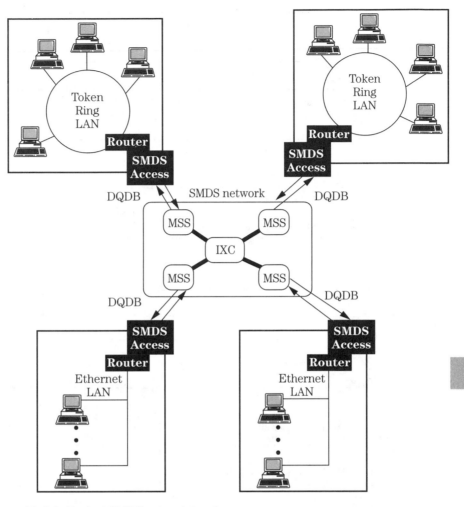

■ **9-9** *Typical SMDS network topology*

SMDS has three different levels of interface to the network, called SMDS Interface Protocols (SIP) (Fig. 9-10). The first, level 3, is a generic datagram service, in which the endpoint equipment pumps data packets of up to 9188 bytes into the SMDS access unit, specifying a unique SMDS address for each datagram. Datagrams are connectionless—that is, no virtual circuit need be established before transmitting data.

In the second SIP, level 2, datagrams are broken into 53-byte chunks called cells. Figure 9-11 illustrates the SMDS cell format. The payload of an SMDS cell includes additional information used by SMDS for the adaptation layer, which assembles and disassembles datagrams; this information occupies four bytes, leaving 44

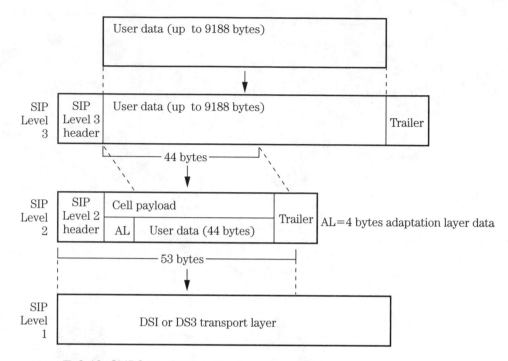

9-10 *SMDS interface protocols*

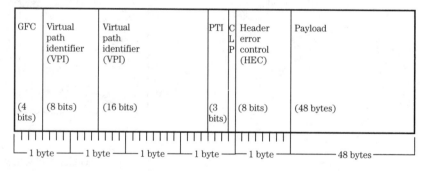

9-11 *SMDS cell format*

bytes for data. Because this adaptation information is encapsulated in the payload field, SMDS cells can move freely across FRS networks.

The third SIP, level 1, is the physical transport protocol. The standard requires this to be DQDB, a transmission system that supports variable data rates up to 45 Mbps easily, and that can be enhanced for even higher rates. Initially, though, SMDS can use DS1 or DS3 lines. Note that these are not T1 or T3 circuits; T, or trunk, circuits have predefined electrical and protocol characteristics. SMDS running over the same physical medium does not follow T1 or T3 conventions, but rather uses its own protocol based on SONET.

An SMDS connection, whether optical or copper, is a bit pipe having variable throughput. To guarantee a certain level of average throughput, SMDS pipes are organized into classes whenever the access speed is higher than 1.544 Mbps. The classes are

Class 1: 4 Mbps
Class 2: 10 Mbps
Class 3: 16 Mbps
Class 4: 25 Mbps
Class 5: 34 Mbps

You can see that classes 1 through 3 match the data rates of Token Ring and Ethernet LANs; these classes are intended to ensure throughput rates required by these networks.

A credit system is used to enforce the throughput limits for each class. The MSS discards traffic from a subscriber when that subscriber's credit balance is insufficient for the class of service required. A node starts out with a certain credit balance, and that balance is debited by an amount based on the number of bytes transferred. As long as the debit would not result in a negative balance, the transmission is accepted. When insufficient credits remain, the node must wait an interval until enough credits accumulate. If a node sends bursty traffic, it will accumulate credits during slack periods, and can then spend them in a burst of high-speed data.

Interconnecting LANs with SMDS

The primary mission in life of SMDS is to interconnect LANs, and to that end it works very well. The interconnections are easy to set up—a single SMDS-capable router can provide all the hardware required—and easy to manage. Because SMDS uses fastpacket technology, it can provide bandwidth-on-demand without complicated bonding techniques, multiple leased channels, or pre-purchased facilities. SMDS, through its access class arrangement, has the ability to carry high speed real-time traffic, including high fidelity audio and video, which makes it a useful vehicle for future applications that need these services.

Glossary

10BASE-2 Standard coaxial cable (e.g., RG-58A/U) for use in Ethernet and IEEE 802.3 networks. An individual 10BASE-2 segment can be up to 185 meters long. Thin cable uses twist-on BNC connectors to attach to a system or other LAN devices.

10BASE-5 A thick (10-mm diameter) coaxial cable used in Ethernet and IEEE 802.3 networks. An individual 10BASE-5 segment can be up to 500 meters long. Thick cables are usually employed as a main LAN backbone.

10BASE-T Unshielded Twisted Pair (UTP) cable suitable for use in Ethernet, Token Ring, IEEE 802.3, or IEEE 802.5 networks. *See also* UTP.

100BASE-T4 A "fast Ethernet" proposal for the use of the standard Ethernet CSMA/CD access discipline over four pairs of type 3, 4, or 5 UTP cable at speeds up to 100Mbps.

100BASE-TX A "fast Ethernet" proposal for use of the standard Ethernet CSMA/CD access discipline over two pairs of type 5 UTP cable at speeds up to 100 Mbps.

100BASE-VG A "fast Ethernet" proposal for the use of a CSMA/CD alternative access discipline over type 3, 4, and 5 UTP cables at speeds up to 100 Mbps. 100BASE-VG is also available in a token-passing variation under the term "100BASE-VG/AnyLan."

100BASE-X A "fast Ethernet" proposal for use of the standard Ethernet CSMA/CD access discipline over type 5 UTP cable at speeds of 100 Mbps.

802.2 LAN communication specification. *See* IEEE 802.2.

802.3 CSMA/CD LAN specification. *See* IEEE 802.3.

802.4 Token bus LAN specification. *See* IEEE 802.4.

802.5 Token Ring LAN specification. *See* IEEE 802.5.

access discipline The low-level protocol used in LANs to control when systems can send and receive information. *See also* CSMA/CD and token passing.

ACCUNET AT&T's digital service providing DDS (2400 bps to 56 kbps), Switched-56 (56 kbps), T1 (1.544 Mbps), and T3 (44.736 Mbps) service.

ACK ACKnowledgment. A control character transmitted by a receiver as an affirmative response to the sender.

ACU Automatic Call Unit. A device used in conjunction with a standard modem that dials the telephone number for the originating equipment. Dedicated ACU devices were popular (and necessary) prior to the advent of Hayes and Hayes-compatible modems. ACUs work in conjunction with asynchronous or synchronous modems.

ADCCP Advanced Data Communications Control Protocol. The ANSI implementation of a bit-oriented, symmetrical protocol based on IBM SDLC. Because of the ANSI endorsement, support for ADCCP is often specified in connectivity situations involving the United States Government.

address A set of bits (or bytes) that uniquely identifies a device on a multidrop (or multipoint) data communications line or in a network.

220

analog transmission Transmission in which the native discrete-valued data-processing digital signals are converted into continuous-valued waveforms for transmission. This is the type of transmission used when sending information over voice-grade phone lines. *See also* digital.

ANSI American National Standards Institute. ANSI is a non-profit, nongovernmental body supported by over 1000 trade organizations, professional societies, and companies. ANSI is the American representative at ISO. *See also* ISO.

APPC Advanced Program-to-Program Communications. An IBM term for an interface that allows two programs running on separate systems to communicate with one another. In most cases, APPC refers to the LU 6.2 interface.

AppleTalk Apple's local area network products and services.

APPN Advanced Peer-to-Peer Networking. An IBM term that refers to the ability of "intelligent" midrange or micro systems in a network to communicate with one another and route messages for one another without involving any higher level SNA devices.

ARCnet A LAN implementation developed by Datapoint that uses a token-passing discipline operating over a 2.5-Mbps physical network. ARCnet became popular in implementing PC networks because it offered a reasonable level of performance at a reasonable price. ARCnet can be implemented in many topologies, but it is usually implemented as a star.

ARP Address Resolution Protocol. A protocol in the TCP/IP suite that translates the logical IP address for a system into the hardware address assigned to the adapter in that system. *See also* RARP.

ARPA Advanced Research Projects Agency. *See also* DARPA.

ARPAnet The first packet-switched network, which offered datagram services in which packets were delivered without regard to sequence or routing.

ASCII American Standard Code for Information Interchange. An ANSI-defined code that defines the bit composition of characters and symbols. ASCII defines 128 different symbols using 7 binary bits (the eighth bit is reserved for parity). DEC, HP, Sun and others use the ASCII encoding system. *See also* EBCDIC.

asynchronous A data transmission method in which each transmitted character (eight bits) is bounded by a start bit and one or more stop bits. Under asynchronous communications, no timing or clocking information is exchanged between parties. *See also* synchronous.

ATM Asynchronous Transfer Mode. A highly efficient cell-switching protocol used to transfer voice, video, and data at high rates over local area and wide area fiber optic networks.

AUI Attachment Unit Interface. The interface on a LAN between a network device (e.g., workstation or computer) and a Medium Attachment Unit. Often used as a qualifying adjective referring to cable (i.e., the "AUI cable" attaches the workstation to the MAU). *See also* MAU and transceiver.

backbone The main cable of a bus or tree LAN to which nodes or other LAN segments are attached.

balanced A balanced circuit uses two wires of instantaneously opposite electrical polarity to carry a data signal. Thus, a simple transmit and receive circuit must contain four wires. As a general rule, balanced circuits are more reliable and accommodate higher transmission rates than unbalanced circuits. *See also* unbalanced.

bandwidth In a digital communications channel, the information-carrying capacity in bits per second.

bandwidth on demand A feature of data communications networks in which bursts of traffic are accepted and transmitted by using alternative communications channels—either additional digital links of the same type, or special digital links that have the capacity to handle high traffic volumes for short periods.

baseband A data communications scheme (often implemented in coaxial cable) in which transmitted digital signals are represented by simple "on" and "off" pulses, rather than being used to modulate a much higher-frequency carrier signal. Baseband signals are used to carry data in many local area networks. *See also* broadband.

baud A unit of data-transfer rate expressed in samples per second. Although baud is often used interchangeably with bits per second (bps), the two units are not necessarily equivalent. While a measurement in bps always specifies bits per second, a measurement in baud does not. If the sampling resolution is one bit, then 1 baud will be the same as 1 bps. If, however, the sampling resolution is two bits, then 1 baud and 1 bps will differ.

BCC Block Check Character. The result of a transmission verification algorithm performed on the block of data being transmitted. The one- or two-character result is normally appended to the end of the transmission. *See also* CRC and LRC.

beacon A special type of network message in a token ring network. A beacon is generated by the first system that detects a serious fault. The beacon message identifies the system reporting the problem and also identifies the address of its Nearest Active Upstream Neighbor (NAUN). *See also* NAUN.

Bell 103 AT&T modem that either originates or answers phone transmissions using asynchronous communications at speeds up to 300 bps.

Bell 113 Same as Bell 103, except that the 113 modem can only originate or only answer (and not automatically switch between answer and originate).

Bell 201 AT&T modem providing synchronous data transmission at up to 2400 bps. The 201 Model B is for leased lines, and the 201 Model C is for dial-up lines.

Bell 202   An AT&T modem that provides asynchronous transmission at speeds up to 1800 bps over a four-wire leased line, or speeds up to 1200 bps using two-wire leased or dial-up lines.

Bell 208   AT&T modem providing synchronous data transmission at up to 4800 bps. The 208 Model A is for leased lines, and the 208 Model B is for dial-up lines.

Bell 209   AT&T modem providing synchronous data transmission over four wires at speeds up to 9600 bps.

Bell 212   An AT&T modem that provides asynchronous or synchronous transmission over leased or dial-up lines at speeds up to 1200 bps. The standard 212 model provides synchronous operation, and the 212 Model A provides asynchronous operation. *See also* V.22.

bis   The suffix *bis* is often used in CCITT recommendations to indicate the second version of a standard (*bis* is Latin for *second*).

B-ISDN   *See* Broadband ISDN.

BNC   Bayonet-Neil-Concelman. A type of twist-on connector used to attach 10BASE-2 (thin coax) cable to systems in an Ethernet or 802.3 network. The bayonet-style connector is named after its originators (Neil and Concelman).

223

BRI   Basic Rate Interface. The low-end interface of the Integrated Services Data Network (ISDN) that offers two 64-kbps data/voice lines and a third 16-kbps management circuit. *See also* ISDN and PRI.

bridge   Bridges are simple networking devices that interconnect two or more LANs. Bridges operate at the lowest network level and are not aware of what networking protocols are in use. *See also* router.

broadband   A high-bandwidth data communications scheme (used, for instance, in CATV cable) capable of transmitting voice, video, and data simultaneously. Information is transmitted by appropriately modulating a high-frequency carrier signal. *See also* baseband.

Broadband ISDN (B-ISDN)   an ITU recommendation that defines data, voice, and video protocols operating in the megabit- to gigabit-per-second range. B-ISDN is a successor to ISDN, which accommodates only speeds up to 1.544 Mbps.

broadcast message A message sent to all systems in a network. Broadcast messages normally use a special destination address (e.g., FF-FF-FF-FF-FF-FF on a LAN) so all systems can recognize it.

brouter A device that includes the functions and features of both a bridge and a router. *See also* bridge and router.

BSC Binary Synchronous Communication. A byte-oriented protocol using synchronous transmission. BSC was widely used by IBM prior to its transition to the bit-oriented SDLC protocol.

burst mode protocol A variation of the NetWare Core Protocol (NCP) used in Novell NetWare networks. Burst mode enables a client to request and receive more data in a single message than under NCP. *See also* NCP.

bus topology A LAN topology that features a linear backbone to which nodes are connected. *See also* tree, ring, and star.

Carrier Sensing The aspect of the CSMA/CD access protocol used in Ethernet and IEEE 802.3 LANs in which a system listens (senses) for data (carrier) on the LAN medium before attempting to transmit.

CBEMA Computer and Business Equipment Manufacturers Association. An association of United States manufacturers that, among other things, sponsors the X.3 standards committee of ANSI. *See also* ANSI and X.3.

CBX Computerized Branch eXchange. A telephone routing exchange driven by an intelligent device (i.e., a computer).

CCIR Consultative Committee for International Radio. An international standards body that sets the rules and requirements for radio communications. CCIR is a committee within the International Telecommunications Union (ITU). *See also* ITU and CCITT.

CCITT Consultative Committee for International Telegraphy and Telephony. An international standards body that sets the rules and requirements for international communications. CCITT is a committee within the larger International Telecommunications Union (ITU), and is best known for the development of the X.25 standard for public data networks. *See also* CCIR and ITU.

CD Carrier Detect. A lead in the RS-232C interface that signals that information is being received over the data link. CD is also sometimes called Data Carrier Detect (DCD). In the full 25-pin

RS-232C standard, CD is pin 8. In the abbreviated 9-bin PC interface, CD is pin 1. *See also* CTS, RS-232C, RTS.

cell   A small unit of information containing data and routing information that is switched through an ATM network.

Cell Relay Service (CRS)   A broadband networking technology in which data to be transmitted is broken into small (53-byte) fixed-length packets and routed over a high speed mesh network. CRS supports real-time applications, such as voice and video, where delays are unacceptable.

CBR   Committed Burst Rate. In Frame Relay Service, the guaranteed maximum data rate of a circuit.

CBS   Committed Burst Size. In Frame Relay Service, the guaranteed maximum number of bits to transfer at the committed burst rate (CBR).

CIR   Committed Information Rate. In Frame Relay Service, the guaranteed average data rate of a circuit.

CMIP   Common Management Information Protocol. CMIP provides the framework for systems to report problems, configuration information, and performance data to a central network management location. CMIP is part of the Open System Interconnect (OSI) Model. *See also* OSI and SNMP.

coax (coaxial cable)   In general, a cabling system that uses a single conductive core surrounded by an insulating medium that is, in turn, surrounded by a conductive sheathing concentric with the core. Coaxial cable is used by IBM to connect their 3270 family of workstations. *See also* Twinax.

collision   The phenomenon in Ethernet and IEEE 802.3 networks whereby two or more systems transmit at the same time on the same LAN segment, causing the two transmissions to overlap and become garbled.

collision detection   The mechanism in Ethernet and IEEE 802.3 networks that detects the occurrence of a data collision.

conditioning   Special equipment added to standard analog phone lines to provide filtering in support of less error-prone data transmission. Various levels of conditioning are available at various costs.

core exterior gateway   A TCP/IP router that provides interconnections inside the Internet network. *See also* noncore exterior gateway and interior gateway.

cps   Character per second. A measurement of data communications speed and throughput.

CPT   Collision Presence Test. *See* heartbeat.

CRC   Cyclic Redundancy Check. An error detection scheme in which the Block Check Character (BCC) is derived from dividing all the serialized bits in a block by a predetermined binary number. *See also* BCC and LRC.

CSMA/CD   Carrier Sense Multiple Access with Collision Detection. The LAN discipline (protocol) used by both Ethernet and the IEEE 802.3 standard. Under CSMA/CD, a device that wishes to transmit on the network first "listens" for other activity. If the network is quiet, the device then attempts to transmit. Data collisions are detected and result in both transmitters attempting to retry their transmissions at a later time.

CSU   Channel Service Unit. The interface to a Digital Data Service (DDS) line. The CSU takes data off the DDS line and feeds it to a Data Service Unit (DSU) that, in turn, interfaces with the terminal or computer equipment. In many cases a CSU is combined with a DSU into a single unit called an Integrated Service Unit (ISU). *See also* DDS, DSU, and ISU.

226

CTS   Clear To Send. A lead in the RS-232C interface. CTS is raised in response to receipt of the Request To Send (RTS) signal. In brief, when one side of the link wishes to transmit, it raises the RTS line. If the other side is ready to receive, it responds by raising the CTS line. Once transmission has begun, the Carrier Detect (CD) line is also raised. In the full 25-pin RS-232C standard, CTS is pin 5. In the abbreviated 9-pin PC interface, CTS is pin 8. *See also* CD, RS-232C, and RTS.

DARPA   Defense Advanced Research Projects Agency. An agency within the United States Department of Defense (DoD) that was instrumental in the development of TCP/IP. *See also* TCP/IP.

datagram   A packet of information in a connectionless network, such as the Internet. Delivery of such packets is not guaranteed, but is only undertaken on a best-effort basis, without regard to sequencing or routing.

DCD   Data Carrier Detect. *See* CD.

DCE   Data Communications Equipment. A device, such as a modem, that facilitates a data communications link. The DCE interfaces with the Data Terminal Equipment (DTE) that is the origin and/or destination of the information. With this in mind, a complete link includes a DTE interfacing with a DCE that interfaces with another DCE that, in turn, interfaces with another DTE. In direct-connection environments, one side of the connection (normally the computer) emulates a DCE interface. *See also* DTE.

DDM Distributed Data Manager. DDM is an IBM LU 6.2-based service that allows access to physical or logical files on remote AS/400 systems.

DDS Digital Data Service. A leased line using digital transmission that can provide data communications rates up to 56 kbps. When DDS is employed, the modems used in conjunction with analog lines are replaced by a Channel Service Unit (CSU) and a Data Service Unit (DSU). *See also* CSU and DSU.

DECnet Digital's line of products that allow communications between DEC systems.

digital transmission Transmission in which information is sent in discrete bit form (i.e., as a stream of 0s and 1s). *See also* analog.

DLSw Data Link Switching. DLSw is a routing protocol designed to "switch" SNA traffic through a TCP/IP network.

DNA Digital Network Architecture. DEC's architecture for the interconnections of its computer and computer-related devices. This is equivalent to IBM's SNA and Sun's ONC.

DoD Department of Defense. *See also* DARPA.

DQDB (Distributed Queue Dual Bus) A token-ring-like access technology for the Metropolitan Area Network (MAN) standard, IEEE 802.6. DQDB is a bus architecture in which bandwidth is divided into timeslots that any attached station can fill with data.

DS0 Digital Service rate 0. The transmission rate (64 kbps) of each of the 24 circuits in a T1 connection. *See also* T1.

DS1 Digital Service rate 1. The combined transmission rate (1.544 Mbps) of all 24 circuits in a T1 connection. *See also* T1.

DSAP Destination Service Access Point. A field in the IEEE 802.2 structure that identifies the target network protocol for a message. *See also* SSAP.

DSPT Display Station Passthrough. DSPT is an IBM LU 6.2-based service that lets a workstation attached to one AS/400 access a second AS/400 as if it were a native workstation.

DSR Data Set Ready. A lead in the RS-232C interface. Data Set Ready is used to signal that the modem (or DCE device) is ready for communications. The counterpart to DSR is Data Terminal Ready (DTR), which is the computer/terminal's signal that it is ready to communicate. In most cases, no communication can take place unless both the DSR and DTR signals are raised. In the full 25-pin RS-232C standard, DSR is pin 6. In the abbreviated 9-pin PC interface, DSR is pin 4. *See also* DTR and RS-232C.

DSU Data Service Unit. A device that interfaces between a Channel Service Unit (CSU) and a terminal or computer. The DSU and CSU work together to interface the computing device to a Digital Data Service (DDS). *See also* CSU, DDS, and ISU.

DTE Data Terminal Equipment. A device such as a terminal or computer that is the origin or destination of information flowing over a data communications link. The DTE interfaces with a Data Communications Equipment (DCE) device (for example, a modem), which handles the actual data communications processing and interfaces with another DCE device (that in turn interfaces with another DTE device). When direct-connect links are used, one side emulates a DCE while the other performs normal DTE functions. *See also* DCE.

DTR Data Terminal Ready. A lead in the RS-232C interface. Data Terminal Ready is used to signal to the modem (or DCE device) that it is ready for communications. The counterpart to DTR is Data Set Ready (DSR), which is the modem's signal that it is ready to communicate. In most cases, no communication can take place unless both the DTR and DSR signals are raised. In dial-in situations, DTR is normally raised when the Ring Indicator (RI) is raised to tell the modem to answer the phone. In the full 25-pin RS-232C standard, DTR is pin 20. In the abbreviated 9-pin PC interface, DTR is pin 4. *See also* DSR, RI, and RS-232C.

duplex *See* half duplex and full duplex.

EBCDIC Extended Binary Coded Decimal Interchange Code. A definition for the bit composition of characters and symbols. EBCDIC uses 256 eight-bit patterns to define 256 different characters, numbers, and symbols. IBM midrange and mainframe systems use the EBCDIC standard. *See also* ASCII.

ECMA European Computer Manufacturers Association. A standards organization composed of Europe-based computer manufacturers. ECMA participates in both CCITT and ISO activities.

EGP   Exterior Gateway Protocol. A protocol used between non-core exterior gateways to interconnect autonomous networks, or between a noncore gateway and a core exterior gateway to connect an autonomous network to the Internet. *See also* core exterior gateway and noncore exterior gateway.

EIA   Electronic Industries Association. A trade organization in the United States specializing in the electrical and functional characteristics of interface equipment. EIA has a close working relationship with ANSI.

EMI   Electromagnetic Interference. Electromagnetic waves that can potentially interfere with the operation of electronic devices. The Federal Communications Commission (FCC) is responsible for regulating how much EMI or Radio Frequency Interference (RFI) an electronic device (such as a computer) is permitted to generate. *See also* FCC and RFI.

encapsulation   A method of moving information for one type of network protocol through a different type of network. Using encapsulation, information for the first network is transported as data by the second network protocol and delivered to a gateway which then extracts the information. *See also* tunneling.

end node   A system or device in a network that cannot forward or reroute packets intended for other nodes. *See also* network node.

Ethernet   A local area network standard that uses the Carrier Sense, Multiple Access with Collision Detection (CSMA/CD) discipline. *See also* IEEE 802.3.

Ethernet switching   A technique for reducing access contention in an Ethernet or IEEE 802.3 LAN by connecting multiple LAN segments to an intelligent, high speed switching hub.

exterior gateway   *See* core exterior gateway and noncore exterior gateway.

fast Ethernet   Ethernet and 802.3 LANs operated at speeds of 100 Mbps. *See also* 100BASE-T4, 100BASE-TX, 100BASE-VG, and 100BASE-X.

fastpacket   A family of networking technologies that use fixed- or variable-length packets in a mesh network to provide point-to-point connectivity between any two stations on the network. A fastpacket network gains its speed by eliminating error detection and correction; such errors must be handled by higher-layer protocols. Frame Relay, Cell Relay, and ATM are all fastpacket networks.

FCC Federal Communications Commission. An agency of the United States government that regulates the use of communications media, such as television waves, radio waves, and other electromagnetic emissions. *See also* EMI and RFI.

FDDI Fiber Distributed Data Interface. FDDI is an ANSI standard for fiber optic networks that employ a token-passing discipline over a ring topology at speeds up to 100 Mbps. Interfaces to traditional LANs are then provided to allow the FDDI network to act as a wide area or metropolitan area network for the LANs attached to it. *See also* MAN and WAN.

FDDI II Fiber Distributed Data Interface, version 2. FDDI-II supports real-time applications by dividing available bandwidth into guaranteed and nonguaranteed parts. The guaranteed bandwidth carries isochronous traffic, such as voice and video, that can't tolerate delivery delays.

FDDI/UTP Fiber Distributed Data Interface over UTP (unshielded twisted pair) wiring. An alternative FDDI cabling specification that permits the use of FTP copper instead of optical fiber, but over significantly shorter distances (100 meters for UTP vs. 2 kilometers for fiber).

fiber optic cable Extruded glass fibers that carry light waves over extended distances, potentially through tortuous curved pathways. Digital signals can be imposed on a light beam—a process called modulation—injected into a fiber optic cable, and recovered from the beam at the other end of the cable.

FIPS Federal Information Processing Specification. Specifications adopted and published by the United States government that are mandated for use by the government and its agencies.

Fractional T1 One or more of the 24 64-kbps channels of a T1 line broken out by the local telephone office and offered to local customers. This enables a range of customers to share the benefits (and cost) of a full T1 line. *See also* T1.

frame A block of information organized in a format appropriate to a specific type of network. For example, in IEEE LANs, data moving across the network is placed in frames structured according to the IEEE 802.2 specifications. *See also* cell and packet.

Frame Relay Service (FRS) A packet-oriented networking technology that supports bandwidths ranging from 56 kbps to 1.544 Mbps. FRS provides permanent virtual circuits (VCs) between any two stations on the same FR network (often called a

Frame Relay Cloud). FRS is a fastpacket network technology, and thus leaves error detection and correction to higher-layer protocols. FRS provides bandwidth-on-demand by guaranteeing a certain committed information rate (CIR) with burst rates up to the committed burst rate (CBR).

FTP   File Transfer Protocol. FTP is a TCP/IP client/server application that enables the transfer of text and binary files between hosts on a TCP/IP network.

full duplex   Simultaneous independent bidirectional transmission.

full-duplex Ethernet   A technique for increasing the throughput of an Ethernet or 802.3 connection by allowing concurrent transmission and reception on independent circuits.

gateway   A device that permits the network activity on one type of network to flow into (or through) another type of network.

GOSIP   Government OSI Profile. A set of requirements issued by the United States (and United Kingdom) government that dictate the use of products within the government and its agencies. *See also* FIPS.

half duplex   Transmission in one of two directions at any given time, but not both directions simultaneously.

HDLC   High-Level Data Link Control. A bit-level protocol for data transmission. HDLC is ISO's implementation of the IBM SDLC standard. HDLC is often used as a high speed, general purpose computer-to-computer link.

heartbeat   A brief signal generated by a transceiver in an Ethernet network after every transmission to inform the adapter that the transmission was free of collisions. The use of heartbeat is optional.

HELLO   The HELLO protocol is an interior gateway protocol used in TCP/IP networks to share routing information. *See also* interior gateway.

hubs   In general, there are two types of hubs. "Dumb" hubs act as LAN segment repeaters or wiring hubs. "Smart" hubs provide sophisticated wiring management, support network management protocols, and often support bridging, routing, and gateway functions.

ICMP   Internet Control Message Protocol. ICMP is responsible for the detection and reporting of link-level errors in TCP/IP networks.

IEEE   Institute of Electrical and Electronics Engineers. A professional society that, among other activities, participates in the development of standards. IEEE recommendations are usually forwarded to ANSI for their endorsement. Among the best-known IEEE standards are their 802.2, 802.3, and 802.5 LAN specifications.

IEEE 802.2 A standard that defines the Logical Link Control (LLC) level of LAN communications. IEEE 802.2 is used in conjunction with the 802.3, 802.4, and 802.5 Medium Access Control (MAC) standards. In terms of layers, 802.2 resides "above" the MAC standards.

IEEE 802.3 A standard that defines the Medium Access Control (MAC) layer for a Carrier Sense Multiple Access with Collision Detection (CSMA/CD) bus network. The IEEE 802.3 standard is not identical to Ethernet, but Ethernet and 802.3 devices can coexist on the same cable. The IEEE 802.3 standard has been adopted by ECMA as ECMA-80, 81, and 82, and by ISO as ISO 8802/3. *See also* Ethernet, ECMA, and ISO.

IEEE 802.4 A standard that defines the Medium Access Control (MAC) layer for a token-passing bus network.

IEEE 802.5 A standard that defines the Medium Access Control (MAC) layer for a token-passing ring network. IBM's implementation of Token Ring conforms to this standard.

IGP Interior Gateway Protocol. Any of the protocols used by interior gateways in a TCP/IP network to share routing information. *See also* interior gateway, HELLO, RIP, and OSPF.

interexchange carriers Long distance service providers; companies that transport voice and data traffic, usually over wideband digital channels on fiber optic cable, between one Local Exchange Carrier (LEC) and another.

interior gateway A TCP/IP router that exists within an autonomous, self-contained network. *See also* core exterior gateway and noncore exterior gateway.

Internet A global web of interconnected computer networks using the TCP/IP protocol for data transport and network management. The Internet grew out of the original ARPAnet, and was for a time a government-funded public network. It is currently being transformed into a commercially-funded public network. Users are usually charged for access to the Internet on the basis of bandwidth rather than usage, although for low-end feeders (e.g., modems up to 28.8 kbps), access may be charged at an hourly rate.

IP Internet Protocol. *See* TCP/IP.

IPX IPX is a multipurpose transport used in Novell NetWare networks. IPX can carry a number of service protocols, including SPX, NCP, SAP, and RIP. *See also* SPX, NCP, SAP, and RIP.

ISDN Integrated Services Data Network. A digital-based network for voice and data lines. From a broader perspective, ISDN is targeted to be an international service for the integration and networking of voice and digital information.

ISO The International Standards Organization is a voluntary, independent organization chartered to define international standards for communications (of all types). ISO is best known for the development of the seven-layer "Basic Reference Model for Open Systems Interconnection," termed the OSI Model. *See also* OSI Reference Model.

ISO 8802/3 ISO's equivalent of the IEEE 802.2 and 802.3 standards for a CSMA/CD LAN. *See also* IEEE 802.2 and IEEE 802.3.

ISO 8802/4 ISO's equivalent of the IEEE 802.2 and 802.4 standards for a token-passing bus LAN. *See also* IEEE 802.2 and IEEE 802.4.

ISO 8802/5 ISO's equivalent of the IEEE 802.2 and 802.5 standards for a token-passing ring LAN. *See also* IEEE 802.2 and IEEE 802.5.

isochronous service Data delivery over a guaranteed channel in real time. Audio and video data are examples of applications that cannot sustain delivery delays and thus require isochronous service. An isochronous service preallocates bandwidth on a channel, whether it is used or not, in order to ensure that the bandwidth is available when needed. FDDI and Cell Relay are two network technologies that support isochronous traffic.

ISU Integrated Service Unit. The combination of a Channel Service Unit (CSU) and Data Service Unit (DSU) into one device. The ISU is used to interface computers and terminals to a Digital Data Service (DDS) line. *See also* CSU, DDS, and DSU.

ITC Independent Telephone Company. A local telephone company that is not a Regional Bell Operating Company (RBOC).

ITU International Telecommunications Union. The ITU is an agency of the United Nations charged to define standards for international telecommunications. The CCITT is a committee of the ITU. *See also* CCITT.

233

jabber A jabber function allows a transceiver in an Ethernet or IEEE 802.3 network to discontinue transmission if the frame being transmitted exceeds the specified limit (1518 bytes).

kbps Kilobits per second. 1024 bits per second (approximately 128 bytes per second).

kBps Kilobytes per second. 1024 bytes per second (approximately 8,192 bits per second).

Kermit A file transfer protocol developed by Columbia University (of New York) and often used to transfer files between PCs and midrange computers. In most implementations, Kermit also includes terminal emulation. Kermit was indeed named after the Muppet.

LAN Local Area Network. A communications architecture that passes information between multiple systems over relatively short distances at very high speeds.

LAN Manager A PC LAN product by Microsoft that provides networking services similar to Novell NetWare. LAN Manager, however, uses the NetBIOS interface instead of Novell's IPX interface. *See also* IPX, NetBIOS, and NetWare.

LAN Server A PC LAN product by IBM that provides networking services similar to Novell NetWare; however, LAN Server uses the NETBIOS interface instead of Novell's IPX interface. *See also* IPX, NETBIOS, and NetWare.

LAP-B Link Access Balanced, revision B. CCITT's implementation of a balanced, bit-oriented protocol based on the IBM SDLC standard. LAP-B is most widely known as the protocol of choice to connect a computer to a packet switching X.25 network.

LAT Local Area Transport. An Ethernet-based DEC protocol implemented for terminal servers to enable terminals connected to a server to establish logical sessions on DEC host nodes without needing to connect through intervening host nodes. *See also* terminal server.

LATA Local Access and Transport Area. Local Exchange Carriers (LECs) operate within a specific geographical region, divided into service areas called LATAs. Calls within a LATA, even when charged a toll, do not use an Interexchange Carrier, or long-distance company.

LAU Lobe Access Unit. LAUs are used in Token Ring LANs to accommodate the connection of two (or more) systems to a single lobe. LAUs can often be chained together to provide additional attachments. *See also* lobe and MAU.

leased line A permanent circuit provided by the telephone company (or similar organization). A leased line may be a direct point-to-point connection, or it may be a multipoint connection. Leased lines are available for either analog (voice grade) or digital transmission. Analog lines may also be conditioned to reduce errors. *See also* conditioning.

LEC Local Exchange Carrier. A local telephone company, which may be a Regional Bell Operating Company (RBOC) such as Pacific Bell, or an Independent Telephone Company (ITC) such as General Telephone (GTE).

LLC Logical Link Control. The highest layer of LAN communications. In the IEEE standards, 802.2 is the LLC. In Ethernet, the LLC and the Medium Access Control (MAC) are intertwined. In terms of the OSI Reference Model, LLC operates at the Data Link level (layer 2). *See also* MAC and OSI Reference Model.

lobe A section of cable in a Token Ring LAN.

LocalTalk Apple's cabling for the AppleTalk LAN.

Logical Unit *See* LU.

LRC Longitudinal Redundancy Check. An error detection scheme in which a check character is generated on the basis of the exclusive or (XOR) of all characters in the block. *See also* BCC and CRC.

LSP LAN Support Programs. IBM-supplied device drivers for network adapters in PCs.

LU Logical Unit. Logical units are part of the IBM SNA structure and correspond to entities (e.g., users, programs, etc.) that request or transmit information through the network.

LU type 0 (LU.T0) Direct-link communication.

LU type 1 (LU.T1) Data processing workstation communication.

LU type 2 (LU.T2) 3270-type workstation communication.

LU type 3 (LU.T3) 3270-type printer communication.

LU type 4 (LU.T4) Word processing workstation communication. Also used for 5250-family printers.

LU type 6.1 (LU.T6.1) Program-to-program communication using one of the following SNA data stream formats: character string, 3270, logical messages services, or user-defined.

LU type 6.2 (LU.T6.2) Program-to-program communication using either the SNA general data stream format or a user-defined data stream.

LU type 7 (LU.T7) 5250-type workstation communication.

MAC Medium Access Control. The lower level of LAN communications concerned with the discipline and topology of the LAN. MAC corresponds to the Physical layer (layer 1) of the OSI Reference Model. *See also* LLC and OSI Reference Model.

MAN Metropolitan Area Network. A somewhat smaller, special-case implementation of a Wide Area Network. MANs typically use fiber optic transmissions to provide high speed communications over relatively small distances (but distances that are beyond the range of a traditional LAN). *See also* FDDI and WAN.

MAP Manufacturing Automation Protocol. A transport system defined by the manufacturing industry to accommodate its specific needs and requirements.

MAU Medium Attachment Unit. A device that physically attaches to a LAN to permit the connection of one or more devices (or LAN segments) to that LAN via an AUI. *See also* transceiver and AUI.

MAU Multistation Access Unit. The central point of connection in a Token Ring network. The MAU is the "ring" aspect of a Token Ring network. Multiple MAUs can be interconnected to increase the size of a ring. *See also* LAU.

Mbps Megabits per second. 1,048,576 bits per second (approximately 131,072 bytes per second).

MBps Megabytes per second. 1,048,576 bytes per second (approximately 8,388,608 bits per second).

MHz One million cycles per second.

microsecond One millionth of a second.

millisecond One thousandth of a second.

MNP Microcom Networking Protocol. MNP is a set of modem protocols used for error control and compression. MNP levels 1 through 4 provide error detection and correction, and level 5 provides a 2:1 data compression ratio. MNP is compatible with the V.42 and V.42bis recommendations.

modem MOdulator-DEModulator. A device that converts between the digital data format used by computers and the analog signals transmitted over a telephone circuit.

multidrop A connection that enables multiple devices to share one physical line. This is often used in the context of a data communications line that has multiple terminals attached to it. *See also* multipoint.

multipoint A connection between multiple devices that enables all attached devices to share a common link. This is often used when a leased telephone line is shared among three or more points. *See also* multidrop.

multiport repeater A hub that enables the connection of systems or additional LAN segments to a central point in an Ethernet or IEEE 802.3 LAN.

NAK Negative AcKnowledgement. A transmission control character transmitted by a receiver as a negative response to the sender. The normal response to a NAK is to retransmit the previous sequence.

nanosecond One billionth of a second.

NAUN Nearest Active Upstream Neighbor. The system in a token-passing LAN that logically precedes a given system. Each system has a different NAUN.

NCP NetWare Core Protocol. The NCP protocol operates beneath the IPX protocol in Novell NetWare networks and handles the mainstream NetWare services, including accessing files and printers on NetWare servers. *See also* IPX.

NetBEUI NETBIOS Extended User Interface. An enhanced version of NetBIOS. *See also* NetBIOS.

NetBIOS NETwork Basic Input Output System. A low-level interface developed by IBM for PCs, primarily to enable them to share files over a network. In brief, NetBIOS enables a PC program to access network-based files, printers, and communicate with other programs. NetBIOS is often emulated by other network systems.

NetWare A PC LAN product by Novell that provides networking services such as file sharing, printer sharing, and application sharing. *See also* IPX.

network node A system or device in a network that has routing capability. *See also* end node.

NFS Network File Services. NFS is used in UNIX networks to facilitate the sharing of directories and files over the network. NFS is similar to the file sharing services offered in a Novell NetWare network.

node *See* end node and network node.

noncore exterior gateway A TCP/IP router that provides a connection between two or more autonomous networks or between an autonomous network and the Internet.

NRZ Non-Return to Zero. A line transmission scheme in which multiple contiguous 1 bits are sent as pulses of opposite polarity, instead of as pulses of the same amplitude separated by a return to the zero (base) line. NRZ is frequently used by IBM equipment and infrequently used by other vendors' equipment.

ONC Open Network Computing. Sun's architecture for the interconnections of its computer and computer-related devices. This is equivalent to IBM's SNA and DEC's DNA.

OSF Open Software Foundation. A group of computer-related companies that came together to define an industry-wide standard for the UNIX operating system and related services.

OSI Reference Model Open Systems Interconnection Reference Model. A layered architecture for the design and implementation of standards that relate to the interconnection of computer systems. The OSI Model is carved into seven layers. These seven layers, from top to bottom, are:

Layer 7: Application (end user and programming services)
Layer 6: Presentation (data conversions and transformations)
Layer 5: Session (logical link setup and management)
Layer 4: Transport (delivery and delivery acknowledgment)
Layer 3: Network (route management)
Layer 2: Data Link (data packaging and transmission)
Layer 1: Physical (physical transmission media)

OSPF Open Shortest Path First. OSPF is a routing protocol developed to replace RIP in the TCP/IP environment. *See also* RIP.

OUI Organizationally Unique Identifier. The first three bytes in an Ethernet or IEEE 802.3 address. The IEEE assigns each manufacturer an OUI value, or a range of values, that are then used as a prefix for their network adapters.

packet A unit of information that contains data, origin information, and destination information, which is switched as a whole through a network. For example, data flowing through an X.25 network is typically carved into 128-byte packets. *See also* cell and frame.

packet switching A networking technique in which multiple devices convert information into smaller packets and then send them on a common network. Within the network, packets may be routed or rerouted through many different nodes, as determined by the network. Packet switching networks can be more cost-effective than leased or dialable networks because charges are typically based on the volume of data instead of connect time or distance.

PAD Packet Assembly/Disassembly. A device that converts between the X.25 packet protocol (normally LAP-B) and a non-packet protocol (such as those used by asynchronous terminals), so that nonpacket devices may use a packet network. In addition to the protocol conversion, the PAD takes large blocks of information from the local device and chops it up into smaller packets for the network. Conversely, the PAD also takes the small packets from the network and assembles them into a large block of information for the local device.

parallel A data transmission method in which the bits in a character are all sent at the same time over eight channels, rather than one after another over a single channel. *See also* serial.

parity The addition of a nondata bit to a byte, making the number of ones in a byte either always odd or always even. This permits the detection of errors in bytes that have single-bit errors.

PBX Private Branch eXchange. A privately owned telephone routing system.

PDN Public Data Network. A telephone company that offers data services to the public. A Public Data Network does NOT have to use packet switching. *See also* PSN.

Physical Unit *See* PU.

point-to-point A physical connection between two, and only two, terminals/computers.

polling A method for a master device to track the status of its attached devices. When the master device polls its attached devices, each device has an opportunity to respond, indicating that it is present.

PPP Point-to-Point Protocol. PPP is a protocol used to transport other protocols between two systems over a serial link. PPP is

239

most commonly used as a means of connecting remote systems into a TCP/IP network. *See also* SLIP.

PRI Primary Rate Interface. The high-end interface in the Integrated Services Data Network (ISDN). The PRI offers 23 64-kbps data/voice lines and a twenty-fourth 64-kbps management line. *See also* BRI and ISDN.

protocol A set of rules whereby two or more devices agree on information and code structures required for successful and error-free communications.

PSDN Packet Switching Data Network. A data network offering packet switching data services. *See also* PSN.

PSPDN Packet Switching Public Data Network. A Public Data Network offering packet switching data services. *See also* PDN and PSN.

PSN Packet Switching Network. A Packet Switching Network routes small fragments of information (called packets) over a series of switched circuits. *See also* X.25.

PSN Packet Switching Node. A Packet Switching Node is a device within a Packet Switching Network that is capable of routing a packet between several other Packet Switching Nodes. *See also* X.25.

PU Physical Unit. A Physical Unit controls the attached links and resources (Logical Units) of a node. *See also* LU.

PU type 1 (PU.T1) Workstations (e.g., 3270 and 5250).

PU type 2 (PU.T2) Cluster controllers (e.g., 3274) and mid-range processors (e.g., System/3X).

PU type 2.1 (PU.T2.1) Midrange processors (e.g., System/3X) that are capable of directly communicating with one another.

PU type 4 (PU.T4) Communications controllers (e.g., 3705).

PU type 5 (PU.T5) Mainframe hosts.

PVC Permanent Virtual Circuit. In a packet switching network, a circuit between two predefined stations. Usually, the physical connection between stations is continuous, although the route packets take may change to accommodate network traffic patterns and to recover from network facility failures. Contrast with a *switched virtual circuit* (SVC).

RARP Reverse Address Resolution Protocol. A protocol in the TCP/IP suite that translates a hardware address assigned to a network adapter into its corresponding logical IP address. *See also* ARP.

RBOC Regional Bell Operating Company. In 1984, the monolithic AT&T Bell Telephone Company was broken up into 22 independent regional companies. Each RBOC operates within a specific geographical region, which is divided into service areas called Local Access and Transport Areas (LATAs). LATAs establish the boundaries between local and long distance services, the latter being the domain of Interexchange Carriers.

RD Receive Data. A lead in the RS-232C interface. RD is used as the reception line for incoming information. In the full 25-pin RS-232C standard, RD is pin 3. In the abbreviated 9-pin PC interface, RD is pin 2. *See also* RS-232C and TD.

repeater A repeater joins two Ethernet or IEEE 802.3 LAN segments to form a single logical segment. *See also* Multiport repeater.

RFI Radio Frequency Interference. Radio waves that can potentially interfere with the operation of electronic devices. The Federal Communications Commission (FCC) is responsible for setting acceptable levels of RFI and Electromagnetic Interference (EMI) generated by electronic devices (such as a computer). *See also* EMI and FCC.

RI Ring Indicator. A lead in the RS-232C interface. RI is used to tell the terminal or computer that the phone is ringing. Normally RI is used in conjunction with dial-up modems, in which case the receiving device can decide if it wants to answer the call or not. In the full 25-pin RS-232C standard, RI is pin 22. In the abbreviated 9-pin PC interface, RI is pin 9. *See also* RS-232C.

RI Ring In. The connection between MAUs in a Token Ring network that brings the ring into a secondary MAU. An RI connection must be made in conjunction with a Ring Out (RO) connection. *See also* RO.

ring topology A LAN topology that features a central ring to which nodes are connected. *See also* bus, tree, and star.

RIP Routing Information Protocol. Variations of the RIP protocol are used in Novell NetWare, TCP/IP, and XNS networks to enable routers to share routing information with one another and with client and server systems.

241

RJE   Remote Job Entry. A protocol for exchanging jobs, printed output, and data between an IBM mainframe and a remote system or RJE workstation.

RO   Ring Out. The connection between MAUs in a Token Ring network that extends the ring from a primary MAU. An RO connection must be made in conjunction with a Ring In (RI) connection. *See also* RI.

routers   Routers are sophisticated networking devices that interconnect and regulate traffic flow between two or more LANs. Routers operate at the network protocol level and can distinguish one protocol from another. *See also* bridge.

RS-232C   An EIA standard for computer/terminal interfaces that defines the electrical and mechanical characteristics for the interconnection of data terminal equipment to data communications equipment. Two implementations of the EIA Registered Standard (RS) 232 remain in use: the C and D versions. Both versions use the same physical 25-pin connector, implement an unbalanced circuit, and can be used for either asynchronous or synchronous communications at speeds up to 20,000 bps. The D version was implemented in 1987 to achieve compatibility with the international V.24 and V.28 recommendations. This means that an interface labeled RS-232C is not fully compatible with V.24 equipment, especially V.24-compliant modems. In a PC environment (and others), it is often the case that only the following nine of the full set of 25 leads are used and available: Carrier Detect (CD), Receive Data (RD), Transmit Data (TD), Data Terminal Ready (DTR), Signal Ground (SG), Data Set Ready (DSR), Request To Send (RTS), Clear To Send (CTS), and Ring Indicator (RI). *See also* CD, RD, TD, DTR, SG, DSR, RTS, CTS, and RI.

RS-422   An EIA standard that describes the electrical characteristics for a balanced circuit that can be used at data rates up to 10 Mbps. RS-422 is not associated with any particular physical interface; however, it's most frequently implemented using the same physical 25-pin connector used for RS-232 connections. RS-422 is compatible with the V.11 recommendation.

RS-449   An EIA standard for computer/terminal interfaces that defines the electrical and mechanical characteristics for the interconnection of data terminal equipment to data communications equipment for use at signalling rates up to 2,000,000 bps. RS-449 is frequently associated with 37-pin and 9-pin connectors.

RTS Request To Send. A lead in the RS-232C interface. RTS is raised by one side of the link when it wishes to transmit. If the other side is ready to receive, it responds by raising the Clear To Send (CTS) line. Once transmission has begun, the Carrier Detect (CD) line is also raised. In the full 25-pin RS-232C standard, RTS is pin 4. In the abbreviated 9-pin PC interface, RTS is pin 7. *See also* CD, CTS, and RS-232C.

SAA Systems Applications Architecture. A set of routines and transport mechanisms developed by IBM to isolate the development of applications from the specifics of the system where they will operate. Under SAA, a program can be developed on one type of IBM system and then easily moved to another type of IBM system. An SAA-compliant application uses three SAA-defined interfaces: the common user interface, the common programming interface, and the common communications support.

SAP Service Access Point. In IEEE 802.2-based networks (e.g., 802.3 and 802.5), the SAP identifies a type of network protocol.

SAP Service Advertising Protocol. In Novell NetWare networks, SAP is a protocol used by file, print, communications, and other types of servers to announce services available on the LAN.

SCSI Small Computer Systems Interface. SCSI is an intelligent, bus-oriented interface that enables a computer to transfer data between a disk, tape, or another computer. A SCSI subsystem can support up to seven devices, and all of these devices can communicate directly with one another. Thus, under SCSI, a computer can request the hard disk to back up to tape, and the hard disk will transfer directly to the tape without any further interaction from the computer.

SDLC Synchronous Data Link Control. A bit-oriented protocol developed by IBM for use in SNA networks. SDLC is the de facto replacement for IBM bisynchronous communications protocols. IBM submitted SDLC to various standards organizations, where it has been adopted as ADCCP, HDLC, and LAP-B.

segment A section of cable in an Ethernet or IEEE 802.3 LAN.

serial A data transmission method in which the bits in a character are transmitted one after another over a single channel.

SG Signal Ground. A lead in the RS-232C interface. Signal Ground provides a common voltage-reference terminal for the two sides of a data communications link. In the full 25-pin RS-232C standard, SG is pin 7. In the abbreviated 9-pin PC interface, SG is pin 5. *See also* RS-232C.

SLIP   Serial Line Internet Protocol. SLIP is a TCP/IP protocol used to connect remote systems into a TCP/IP network. *See also* PPP.

SMDS   Switched Multimegabit Data Service. A high speed data transmission service offered by local phone companies in the United States. SMDS supports Fractional T1, T1, and T3 data rates.

SMTP   Simple Mail Transfer Protocol. SMTP handles the routing of mail in a TCP/IP network.

SNA   Systems Network Architecture. IBM's architecture for the interconnections of its computer and computer-related devices. This is equivalent to DEC's DNA and Sun's ONC.

SNADS   SNA Distribution Services. SNADS is an IBM LU 6.2-based object distribution service that can be used to transport mail, documents, files, and other objects over the network.

SNMP   Simple Network Management Protocol. SNMP provides the framework for systems to report problems, configuration information, and performance data to a central network management location. SNMP originated in TCP/IP networks, but is now used in other network architectures as well. *See also* CMIP.

socket   A program-level interface in TCP/IP networks. Every TCP/IP host has a number of sockets available, most of them preassigned to specific functions (e.g., FTP, TELNET, etc.). All communications between client and server programs in a TCP/IP network occur over sockets.

source routing   A technique that allows a system located on one physical ring to communicate with a system located on a different physical ring.

SPREAD   A routing protocol used by core exterior gateways to share information with one another. This protocol is only used within the Internet. *See also* core exterior gateway.

SPX   Sequenced Packet eXchange. SPX is a "connection-oriented" protocol in Novell NetWare networks that runs as an extension to the IPX protocol and provides confirmation (or denial) of the end-to-end delivery of messages.

SSAP   Source Service Access Point. A field in the IEEE 802.2 structure that identifies the network protocol used to originate a message. *See also* DSAP.

StarLAN A LAN implementation developed by AT&T that uses 1-Mbps cables in a star topology. A modified version of StarLAN was upgraded to 10 Mbps and given the name StarLAN-10.

star topology A LAN topology that features a central hub to which nodes are connected. *See also* bus, tree, and ring.

STP Shielded Twisted Pair. Data communications cable in which connector pairs are twisted together and then enclosed in a protective and shielded outer sheath.

SQE Signal Quality Error. *See* heartbeat.

SVC Switched Virtual Circuit. In a packet switching network, a circuit that can connect, on demand, to any other station on the network. *See also* Permanent Virtual Circuit (PVC).

SW56 Switched-56. A digital data service that transports data over switched synchronous lines at 56 kbps, or over switched asynchronous lines at 57.6 kbps. Calls are carried on the Public Switched Telephone Network (PSTN), and can thus provide cross-country and even international data transport (at long distance telephone rates).

switched line The line connection made as a result of dialing via the phone system (as opposed to a permanent, leased line).

synchronous A form of transmission in which the sender and receiver exchange timing information on separate channels in order to send a frame with no space or marking between characters. Because no start/stop bits are required, synchronous transmission is more efficient than asynchronous transmission for long messages. *See also* asynchronous.

T1 A standard definition for digital transmission in the Bell System T-carrier digital environment. T1 defines a path with 1.544-Mbps bandwidth that can be divided into 24 channels of 64-kbps service. Each of the individual channels is said to have a Digital Signal Level Zero (DS0) rate, and all 24 as a whole result in the Digital Signal Level One (DS1) rate. *See also* Fractional T1.

T3 A standard definition for digital transmission in the Bell System T-carrier digital environment. T3 defines a path with 44.736-Mbps bandwidth that can be divided into 28 channels of 1.544-Mbps service. Each of the individual channels is said to have a Digital Signal Level One (DS1) rate, and all 28 as a whole result in the Digital Signal Level Three (DS3) rate.

TCP/IP Transmission Control Protocol/Internet Protocol. TCP/IP is a set of network services that provide interoperability between heterogeneous systems. The TCP portion is a connection-oriented protocol responsible for providing reliable and recoverable communications between two endpoints. The IP portion handles the routing and delivery of TCP messages. Other protocols that are also part of the TCP/IP suite include UDP (User Datagram Protocol), ICMP (Internet Control Message Protocol), TELNET, FTP (File Transfer Protocol), and SMTP (Simple Mail Transfer Protocol). *See also* FTP, ICMP, SMTP, TELNET, and UDP.

TCP *See* TCP/IP.

TD Transmit Data. A lead in the RS-232C interface. TD is used to transmit information across the interface. In the full 25-pin RS-232C standard, TD is pin 2. In the abbreviated 9-pin PC interface, TD is pin 3. *See also* RS-232C and RD.

TELNET TELNET is a TCP/IP client/server application that allows a TCP/IP workstation or a terminal attached to a TCP/IP host to access a second host system over a TCP/IP network.

ter The suffix *ter* is often used in CCITT recommendations to indicate the third version of a standard (*ter* is Latin for *third*).

terminal server A product that connects terminals to a LAN, enabling the terminals to establish sessions on host nodes. Terminal servers typically implement the Digital LAT protocol, the TCP/IP TELNET service, or both.

ThickLAN A term used to define the standard baseband coaxial cable (10BASE-5) used as the backbone in most IEEE 802.3 and Ethernet LANs.

ThinLAN A term used to define the thin coaxial cable (10BASE-2) used to interface Personal Computers and office equipment in a LAN environment. ThinLAN segments may be interfaced to a ThickLAN backbone (or segment).

ThinWire The DEC term corresponding to ThinLAN.

token A special message used in Token Ring network. The token message is three bytes long, and is a variation of the standard frame format used to carry messages. *See also* token-passing.

token passing discipline Token passing is a LAN discipline in which a specific message, termed a token, is passed from device to device on the LAN. The device that possesses the token has the ability to transmit on the LAN, and when the device is finished transmitting, it releases the token to the next device downstream. Token passing and CSMA/CD are the two most prevalent LAN disciplines. *See also* CSMA/CD.

Token-Passing Ring *See* IEEE 802.5.

Token Ring Network *See* IEEE 802.5.

topology The physical organization and structure of a network. *See* bus, tree, ring, and star.

transceiver The device that attaches a node to a LAN. Transceivers interface with transceiver cables, which attach, in turn, to the nodes. *See also* MAU and AUI.

transparent data Binary data transmitted with the recognition of control characters suppressed.

tree topology A LAN topology that features a linear backbone to which other nodes or LAN segments connect. The tree topology and bus topology are very similar, and the two terms are in fact often interchanged. *See also* bus, ring, and star.

TTY TeleTYpe. Originally a specific keyboard/hardcopy device, the term TTY has gone on to become a general industry term describing a "dumb" terminal that is operated on a key-by-key or line-by-line basis.

tunneling A method of moving information for one type of network protocol through a different type of network. Using tunneling, information for the first network is inserted into a second network at one point and removed from that network at a different point. *See also* encapsulation.

Twinax IBM's twinaxial cabling system for the 5250 family of workstations, as used in conjunction with the IBM midrange system line (e.g., System/36, System/38, and AS/400). Twinax is similar to coax except, as the name implies, it uses two conducting cores. *See also* coax.

UDP User Datagram Protocol. UDP is an alternative transport to TCP. Like TCP, UDP relies on IP to handle the routing and delivery of messages. Unlike TCP, UDP is a connectionless protocol that does not guarantee end-to-end delivery. Because it is connectionless, UDP is faster than TCP and is typically used for real-time client/server applications. *See also* TCP/IP.

unbalanced A simple unbalanced circuit typically contains one wire to transmit and one to receive (and often uses a third ground wire).

UTP Unshielded Twisted Pair. As the name implies, UTP cable is made of pairs of wires twisted together and enclosed in a protective (but unshielded) sheath. Five types of UTP cable are available:

Level 1: Data transmission rates up to 4 Mbps.
Level 2: Data transmission rates up to 4 Mbps.
Level 3: Data transmission rates up to 16 Mbps.
Level 4: Data transmission rates up to 20 Mbps.
Level 5: Data transmission rates up to 100 Mbps.

V.11 A CCITT recommendation for the electrical characteristics of a balanced circuit that supports data rates up to 10 Mbps. V.11 operation is compatible with the RS-422 standard.

V.21 A CCITT recommendation for modem operations that use asynchronous transmission over dial-up lines at speeds up to 300 bps.

V.22 A CCITT recommendation for modem operations that use asynchronous or synchronous transmission over leased or dial-up lines at speeds up to 1200 bps. V.22 is compatible with the Bell 212 specifications.

V.22bis A CCITT recommendation for modem operations that use asynchronous or synchronous transmission over leased or dial-up lines at speeds up to 2400 bps, with a fallback capability of 1200 bps.

V.23 A CCITT recommendation for modem operations that use asynchronous transmission over dial-up lines at speeds up to 1200 bps.

V.24 A CCITT recommendation that defines the physical exchange circuits used to interconnect computer equipment. RS-232D is a subset of the full V.24 recommendation. Note that although V.24 defines the purpose of each circuit, the V.28 circuit defines the electrical properties of the circuit.

V.25 A CCITT recommendation for an automatic calling and/or answering unit that requires two attachment interfaces (two cables). V.25 has fallen out of favor by virtue of the V.25bis recommendation.

V.25bis A CCITT recommendation for an automatic calling and/or answering unit. V.25bis requires a single interface (one cable) and implements a command protocol that's conceptually similar to the Hayes AT modem interface, but is specifically oriented toward international phone circuits.

V.27 A CCITT recommendation for modem operations that use synchronous transmission over leased lines at speeds up to 4800 bps.

V.27bis A CCITT recommendation for modem operations that use synchronous transmission over leased lines at speeds up to 4800 bps, with a fallback rate of 2400 bps. V.27bis also implements automatic line equalization to reduce transmission errors.

V.27ter A CCITT recommendation for modem operations that use synchronous transmission over dial-up lines at speeds up to 4800 bps, with a fallback rate of 2400 bps. In a nutshell, V.27ter is a version of V.27bis intended for dial-up lines.

V.28 A CCITT recommendation that defines the electrical characteristics of an unbalanced interchange. V.28 is most commonly used with a V.24 physical circuit. Under V.28, signaling voltages between +5 and +15 volts are defined to be binary 0, and voltages between –5 and 15 volts are defined to be binary 1. An RS-232D circuit shows these same characteristics.

V.32 A CCITT recommendation for modem operations that use asynchronous or synchronous transmission over dial-up lines at speeds of 9600 bps and 4800 bps.

V.32bis A CCITT recommendation for modem operations that use asynchronous or synchronous transmission over dial-up lines at speeds up to 14,400 bps, with a fallback rate that's negotiated to the highest available speed.

V.32ter AT&T's revision of the CCITT V.32 standard, V.32ter defines modem operations that use asynchronous or synchronous transmission over leased or dial-up lines at speeds up to 19,200 bps, with a fallback rate that's negotiated to the highest available speed.

V.33 A CCITT recommendation for modem operations that use synchronous transmission over four-wire leased lines at speeds up to 14,400 bps, with a fallback rate of 12,000 bps.

V.35 A CCITT recommendation that defines data transmission at speeds up to 48 kbps over modems and DSUs. V.35 is actually an obsolete recommendation that has been replaced by the V.36 and V.37 recommendations. However, because V.36 and V.37 continue to use the same physical circuit defined by V.35, the computer industry continues to use V.35 as a "generic" term. Therefore V.35 is often used for speeds of 56 kbps, 64 kbps, and higher.

V.36 A CCITT recommendation that defines synchronous data transmission at speeds up to 72 kbps over modems and DSUs. *See* V.35.

V.37 A CCITT recommendation that defines a synchronous data transmission circuit that can achieve speeds in excess of 72 kbps over modems and DSUs. *See* V.35.

V.42 A CCITT recommendation derived from MNP that applies error control and minimal data compression to data flowing between two modems. V.42 is a superset of MNP levels 1–4. *See also* V.42bis.

V.42bis A CCITT recommendation that defines a data compression scheme that may be used in conjunction with the V.42 recommendation to achieve a data compression ratio of up to 4:1. V.42bis includes MNP level 5 capabilities.

VAN Value Added Network. A private network (normally a wide area network) offered on a commercial basis. Customers attach to a VAN to access value-added services, or to interconnect their own geographically dispersed equipment.

VINES A PC LAN product by Banyan that provides networking services similar to Novell NetWare.

WAN Wide Area Network. A network composed of systems that are relatively far apart. A WAN may also encompass a series of LANs connected together over a wide area. *See also* MAN.

X.3 A standards committee within ANSI focused on the areas of computers and networking.

X.21 A CCITT recommendation that defines the characteristics of an interface to attach to a public data network. The X.21 recommendation describes a 15-pin interface that conforms to the V.11 electrical interface and supports synchronous transmission at rates up to 10 Mbps. X.21 is an alternative to V.35 for connection to high speed digital networks. *See also* X.21bis.

X.21bis A CCITT recommendation that describes the use of V.24 and V.35 interfaces with public data networks. The X.21bis recommendation defines the behavior of the individual wires within each circuit at all phases of a connection. *See also* X.21.

X.25 A CCITT recommendation that defines the interface between a packet-mode host system and a packet switching network. Among other things, the X.25 recommendation describes the protocol to be used by a packet-mode system that attaches to a packet switching network by means of a permanent (leased) connection. *See also* X.28, X.29, X.32, and X.75.

X.28 A CCITT recommendation that defines an interface between an asynchronous device (e.g., an ASCII terminal) and a PAD within a packet switching network. *See also* X.25.

X.29 A CCITT recommendation that defines the interface between a PAD and a packet-mode host system, or between two PADs. *See also* X.25.

X.32 A CCITT recommendation that defines the interface used by a packet-mode host system that attaches to a packet switching network by means of a switched (i.e., dial-up) connection. In essence, X.32 lets the services described in the X.25 recommendation be carried over a switched connection, such as public telephone or ISDN.

X.75 A CCITT recommendation that describes the protocols and procedures used to exchange information between two packet switching networks. In other words, the X.75 recommendation lets two X.25 packet switching networks be connected together.

X.400 A standard for implementing electronic mail on diverse computer systems. X.400 has provisions for the exchange of messages, files, and video information.

X.500 A standard for implementing common user name/directory services on heterogeneous computer systems.

Xmodem A simple file transfer protocol used to transfer data between two computers. The most typical use for Xmodem is to transfer files between a host computer and a PC. *See also* Kermit and Zmodem.

XNS Xerox Networking System. Xerox Corporation's networking services implemented over an Ethernet LAN.

XON/XOFF A simple pacing mechanism implemented between sending and receiving units. Under this mechanism, the sender transmits until the receiver sends an XOFF character, signaling the transmitter to pause. When the receiver is ready for more data, it sends an XON character, and the transmitter resumes sending. The XON and XOFF characters are normally the DC1 and DC3 control characters, respectively.

Zmodem A file transfer protocol used to transfer data between two computers. Like Xmodem, the typical use of Zmodem is to transfer files between a host computer and a PC. Zmodem was developed after Xmodem and offers better performance. *See also* Kermit and Xmodem.

Index

---

*Illustrations are indicated in **boldface**.*

innerducts, 150
integrated services digital
network (ISDN), 7-9, 176,
179, 188, 198, 207-213,
210, **212**
asynchronous transfer
mode (ATM), 18-19
broadband ISDN (B-ISDN),
18
speed of transmission, 7
Intel Corporation, 22, 81
interactive user environ-
ments, 3
interconnection tools (*see*
bridges; gateways; hubs;
routers)
International Standards Organization (ISO), 21
Internet, 111
Internetwork Packet eX-
change (IPX), 16, 87
interoperability gateways,
103-104, 112-115 **114**
intranetwork gateways, 103
IPX, 26, 74, 75, 78, 83, 84,
89, 103, 104, 142
gateways, 105-106, **106**, 108

J

jabber, Ethernet/802.3, 37
jamming signals, Ether-
net/802.3, 37

L

LAN management protocol,
26
LAN support program (LSP),
92-93
LAT, 97
lobe access test, Token
Ring/802.5, 55
lobe access unit (LAU), Token Ring/802.5, 62, **63**,
64, **65**
lobe test (phase 0), Token
Ring/802.5, 54
lobes, Token Ring/802.5, 62
local access component
(LAC), 199
local area networks (LAN),
1, 2, 3, 11-15, 21-50, 93
802.X standards, 13

access protocol, 80-83, **82**,
85
address resolution protocol
(ARP), 92
advanced peer-to-peer networking (APPN), 93, 102
advanced program-to-pro-
gram communication
(APPC), 89
ARCnet, 14, 81
asynchronous transfer
mode (ATM), 18-19, 174
attached resource com-
puter network (ARCnet),
12
bandwidth, 16
bridges, 16-17, 77-99, **79**
broadband ISDN (B-ISDN),
18
broadcast addresses, 82-83
brouters, 78, 97
Burst Mode protocol (Net-
Ware), 87
bus topologies, 12, 13
carrier sense multiple access with collision detection (CSMA/CD), 13, 22,
24, 81, 84
choosing LAN implementa-
tion, 26-28
control fields, 25, **25**
data fields, 25, **25**
data link switching (DLSw)
routers, 93-94
destination service access
point (DSAP), 25-26, **25**,
84, 93
display station passthrough
(DSPT), 89
distributed data manager
(DDM), 89
Ethernet/802.3, 12, 13, 14,
15, 21-23, 26, 27, 28, 29-
50, 77, 81, 84, 99, 102
fiber distributed data interface (FDDI), 15, 19, 28,
81, 102, 102
fiber optic networks, 14,
18, 80, 101
file transfer protocol
(FTP), 88
frame relay, 17

frames, 15
future developments/up-
grades, 28
gateways, 16-17, 101-121
global LAN access (802.2),
13
history and development, 12
hubs, 101-121
IBM connections, 27
IEEE standards, 13, 21-22,
23-26
installation ease, 27
internetwork packet ex-
change (IPX), 87
ISDN interconnection, 213
joining (interconnecting)
LANs, 77-80
LAN support program
(LSP), 92-93
logical link control (LLC),
24-25, **24**
media access control
(MAC), 24, **24**
multi-vendor environ-
ments, 27
multiprotocol "super-
servers," 83
multistation access unit
(MAU), 27
NetWare, 87
NetWare Core Protocol
(NCP), 87
network protocols, 81-83,
82, 83-89, **85**
number of stations in network, 27-28
peer-oriented communica-
tion, 16
performance choices, 28
protocol type fields, 84-85,
93
protocols, 26, 80-89
ring topologies, 13
routers, 16-17, 77-99, **79**
routing information protocol (RIP), 87, 88
SAP assignments, IEEE
802.2, 84
sequenced packet ex-
change (SPX), 87
Service Advertising Protocol (SAP), NetWare, 87, 92

259

260

261

263

264

265

About the Authors

John Enck has over 15 years of experience in data communications and networking. He is currently network analyst for Forest Computer, a manufacturer of multivendor gateways. **Mel Beckman** is a specialist in the design, implementation, and application of communication networks and protocols. He has authored several commercial software products and currently serves as senior editor for *NEWS 3X/400*, an IBM systems journal.

Other Bestsellers of Related Interest

McGraw-Hill LAN Communications Handbook
—Fred Simonds
Starting with the basic concepts and progressing to specifics, this handbook helps readers understand what a LAN is and does. It discusses every aspect of LAN hardware and software to help users decide if they need a LAN and, if so, how to go about purchasing, installing, and running one.
0-07-057442-1 $65.00 Hard

SNMP: A Guide to Network Management
—Sidnie Feit
Written for both those who plan, administer, and manage networks and for software developers who work in a networked environment, this reference presents all the ideas behind SNMP and clearly explains the protocols and mechanisms.
0-07-020359-8 $50.00 Hard

Disaster Recovery for LANs: A Planning and Action Guide
—Regis J. "BUD" Bates
Provides all of the facts LAN administrators, network planners, MIS/DP directors, and senior managers need to determine where their exposures lie, what protection will cost, and what their options are.
0-07-004494-5 $34.95 Paper
0-07-004194-6 $45.00 Hard

Beyond LANS: Client/Server Computing
—Dimitris N. Chorafas
Updates, expands, and provides experienced data processing professionals with the knowledge and skills needed to understand and implement the logical extension of LANs—client/server computing.
0-07-011057-3 $50.00 Hard

ISDN: Concepts, Facilities, and Services, Second Edition
—*Gary C. Kessler*
Aimed primarily at telecommunications and data communications specialists designing and implementing ISDN networks, this new edition also features many revised examples, more than 100 clear illustrations, an updated list of acronyms and abbreviations, and a complete glossary for easy reference.
0-07-034247-4 $50.00 Hard

Open Systems Interconnection: Its Architecture and Protocols, Revised Edition
—*B. N. Jain and A. K. Agrawala*
This introduction to OSI fundamentals provides extensive coverage of the services and protocols at each layer, with an emphasis on the upper layers.
0-07-032385-2 $50.00 Hard

LAN TIMES Encyclopedia of Networking
—*Tom Sheldon*
Organized alphabetically, this unparalleled authority on the subject of networking give readers easy-to-comprehend information on the most important topics in the world of local and global networks.
0-07-881965-2 $39.95 Paper

Network Security: How to Plan For It and Achieve It
—*Richard Baker*
An informed and practical guide on how to make networked computers as secure as they can be. Network specialists, administrators, managers, and planners will see how to secure a client/server database; how to tackle remote access problems; how to resolve E-mail security/privacy issues; and more.
0-07-005141-0 $34.95 Paper

Network Management Standards: SNMP, CMIP, TMN, MIBs, and Object Libraries
—*Uyless Black*
All network management standards in one easy-to-read book. A detailed examination of the OSI, SNMP, and CMOL network management standards.
0-07-005570-X $50.00 Hard

Wireless Networked Communications
—Bud Bates

This book is a single source of information for telecommunications and LAN managers, network planners, and others who want a clear explanation of today's wireless communications systems and how to implement them.

0-07-004674-3 $50.00 Hard

Enterprise Network Performance Optimization
—Martin Nemzow

This book is especially helpful for large organizations with overloaded network infrastructures and performance problems, the guide emphasizes the concept of applying true scientific methods to enable readers to best collect data, analyze performance, and improve performance.

0-07-911889-5 $50.00 Hard

Linking LANs-2/e
—Stan Schatt

LAN designers, managers, and administrators concerned with integrating LANs into an enterprise-wide or wide-area network will find the answers to their questions in this comprehensive guide. This updated edition contains new chapters on the enterprise networks, wireless LAN technology, linking branch offices, TCP/IP protocols, asynchronous transfer mode, and frame relays for wide-area networks.

0-07-057063-9 $45.00 Hard

TCP/IP and Related Protocols-2/e
—Uyless Black

This new edition clearly explains all the facets of the Transmission Control Protocol/Internet Protocol and the many protocols that operate within these two standards. Updated coverage explains new resource records for DNS; discovery services and messaging in the Internet.

0-07-005560-2 $50.00 Hard

How to Order

Call 1-800-822-8158
24 hours a day,
7 days a week
in U.S. and Canada

Mail this coupon to:
McGraw-Hill, Inc.
P.O. Box 182067,
Columbus, OH 43218-2607

Fax your order to:
614-759-3644

EMAIL
70007.1531@COMPUSERVE.COM
COMPUSERVE: GO MH

Shipping and Handling Charges

Order Amount	Within U.S.	Outside U.S.
Less than $15	$3.50	$5.50
$15.00 - $24.99	$4.00	$6.00
$25.00 - $49.99	$5.00	$7.00
$50.00 - $74.49	$6.00	$8.00
$75.00 - and up	$7.00	$9.00

EASY ORDER FORM—
SATISFACTION GUARANTEED

Ship to:

Name _____

Address _____

City/State/Zip _____

Daytime Telephone No. _____

Thank you for your order!

ITEM NO.	QUANTITY	AMT.

Method of Payment:

☐ Check or money order
enclosed (payable to
McGraw-Hill)

☐ VISA   ☐ DISCOVER

☐ AMERICAN EXPRESS Card   ☐ MasterCard.

Shipping & Handling charge from chart below	
Subtotal	
Please add applicable state & local sales tax	
TOTAL	

Account No. ☐☐☐☐☐☐☐☐☐☐☐☐☐☐☐☐

Signature _____ Exp. Date _____
Order invalid without signature

**In a hurry? Call 1-800-822-8158 anytime,
day or night, or visit your local bookstore.**

Code = BC15ZZA